# Synthesis Lectures on Renewable Energy Technologies

The series, Synthesis Lectures on Renewable Energy Technologies publishes concise books, focused on technologies that harness energy from naturally occurring sources, such as sunlight, wind, water, geothermal heat, and biofuels from organic materials. These renewable energy technologies play a crucial role in transitioning away from fossil fuels, helping to mitigate the effects of climate change, and promoting a sustainable energy supply.

Mónica L. Chávez-González ·
Pedro Aguilar-Zarate · A. K. Haghi

# Energy Recovery from Agro-Food Waste

New Technologies for Sustainable Development

Mónica L. Chávez-González
School of Chemistry
Universidad Autónoma de Coahuila
Saltillo, Coahuila, Mexico

Pedro Aguilar-Zarate
Tecnológico Nacional de México/Instituto
Tecnológico de Ciudad Valles
Ciudad Valles, San Luis Potosí, Mexico

A. K. Haghi
Institute of Molecular Chemistry
Coimbra University
Coimbra, Portugal

ISSN 2690-5000  ISSN 2690-5019 (electronic)
Synthesis Lectures on Renewable Energy Technologies
ISBN 978-3-031-89098-7  ISBN 978-3-031-89099-4 (eBook)
https://doi.org/10.1007/978-3-031-89099-4

© The Editor(s) (if applicable) and The Author(s), under exclusive license to Springer Nature Switzerland AG 2026

This work is subject to copyright. All rights are solely and exclusively licensed by the Publisher, whether the whole or part of the material is concerned, specifically the rights of translation, reprinting, reuse of illustrations, recitation, broadcasting, reproduction on microfilms or in any other physical way, and transmission or information storage and retrieval, electronic adaptation, computer software, or by similar or dissimilar methodology now known or hereafter developed.
The use of general descriptive names, registered names, trademarks, service marks, etc. in this publication does not imply, even in the absence of a specific statement, that such names are exempt from the relevant protective laws and regulations and therefore free for general use.
The publisher, the authors and the editors are safe to assume that the advice and information in this book are believed to be true and accurate at the date of publication. Neither the publisher nor the authors or the editors give a warranty, expressed or implied, with respect to the material contained herein or for any errors or omissions that may have been made. The publisher remains neutral with regard to jurisdictional claims in published maps and institutional affiliations.

This Springer imprint is published by the registered company Springer Nature Switzerland AG
The registered company address is: Gewerbestrasse 11, 6330 Cham, Switzerland

If disposing of this product, please recycle the paper.

# Contents

**1 Introduction to Food Waste-to-Energy Conversion Technologies** .......... 1
   1.1 Understanding Food Waste-to-Energy Conversion Technologies ........ 1
   1.2 Rationale for Food-Waste-Based Energy ............................ 4
   1.3 Environmental Benefits and Challenges ............................ 7
   1.4 Scope and Structure of the Book .................................. 8
   References .......................................................... 8

**2 Sustainable Approaches and Characteristics** ......................... 11
   2.1 Types of Food Waste-to-Energy Production ......................... 11
   2.2 Composition of Food Waste ....................................... 12
       2.2.1 Pretreatments Used on Food Wastes ........................ 13
   2.3 Food Waste to Energy Research and Development .................... 14
   2.4 Thermochemical Conversion ....................................... 15
   2.5 Anaerobic Digestion ............................................. 16
   2.6 Biochemical Conversion .......................................... 16
       2.6.1 Biorefinery Concept ....................................... 17
       2.6.2 Bioelectricity Generation ................................. 17
   2.7 Factors Affecting Food Waste-to-Energy Conversion ................ 18
   2.8 Thermal Conversion System ....................................... 18
       2.8.1 Ash Melting ............................................... 18
       2.8.2 Corrosion ................................................. 18
       2.8.3 Nitrogen Content .......................................... 18
   2.9 Biochemical Conversion .......................................... 19
       2.9.1 Lignin Content ............................................ 19
       2.9.2 Cellulose and Hemicellulose Content ....................... 19
       2.9.3 Ash Content ............................................... 19
   2.10 Anaerobic Digestion ............................................ 19
   References ........................................................ 20

| 3 | Current Conversion Technologies and Case Studies | | 23 |
|---|---|---|---|
| | 3.1 | Introduction | 23 |
| | 3.2 | Energy Recovery from Food Waste by Combustion | 24 |
| | | 3.2.1 Integration with Circular Economy Principles | 26 |
| | | 3.2.2 Challenges, Limitations and Future Directions | 27 |
| | 3.3 | Biodiesel Production from Agro-Industrial Waste | 28 |
| | 3.4 | Corn Waste to Gasoline | 29 |
| | | 3.4.1 Technological Pathways for Conversion | 30 |
| | | 3.4.2 Sustainability and Environmental Benefits | 32 |
| | | 3.4.3 Economic Viability and Challenges | 32 |
| | 3.5 | Biodiesel Production from Waste Cooking Oil | 33 |
| | | 3.5.1 Advancements in Biodiesel Production Processes | 33 |
| | | 3.5.2 Policy and Regulatory Support | 34 |
| | | 3.5.3 Economic Viability and Applications | 35 |
| | | 3.5.4 Challenges and Research Directions | 35 |
| | References | | 37 |
| 4 | Biogas Production from Food Waste | | 41 |
| | 4.1 | Introduction | 41 |
| | 4.2 | Biotransformation of Food Waste into Biogas | 43 |
| | | 4.2.1 Advances in Biotransformation Techniques and Optimization | 44 |
| | | 4.2.2 Integration with the Circular Economy | 45 |
| | | 4.2.3 Emerging Technologies and Innovations | 47 |
| | 4.3 | Reactor Configuration and Design | 48 |
| | | 4.3.1 Multi-phased and Staged Reactors | 49 |
| | | 4.3.2 Bioelectrochemical Systems (BES) Integration | 51 |
| | | 4.3.3 Batch and Continuous Systems | 51 |
| | | 4.3.4 Temperature Optimized Bioreactors | 53 |
| | | 4.3.5 Fixed-Bed and Packed-Bed Reactors | 54 |
| | 4.4 | Biogas Utilization and Energy Recovery | 56 |
| | | 4.4.1 Innovations in Biogas Production and Upgrading | 56 |
| | | 4.4.2 Recovery and Utilization Pathways | 56 |
| | | 4.4.3 Environmental, Economic, and Policy Impacts | 57 |
| | 4.5 | Future Trends in Biogas Utilization and Technology | 58 |
| | References | | 60 |

# About the Authors

**Mónica L. Chávez-González** is a full researcher/professor at the School of Chemistry of the Universidad Autónoma de Coahuila. She earned her Ph.D. degree in Food Science and Technology. She worked in the development of bioprocesses for the valorization of agroindustrial by-products. She is a member of the Mexican Association of Food Science (AMECA) and a member of the International Bioprocessing Association. She is the General Coordinator and founder of the Bioprocessing Mx-LATAM Network, whose objective is to promote R&D in the area of bioprocessing through interaction between universities, institutes, research centers, and relevant industries in Mexico and Latin America.

Dr. Chávez has been a member of the National System of Researchers (SNII, Level II, Mexico) since 2016. She has published over 65 articles in JCR journals, 61 chapter books, and more than 10 books in prestigious international editorials. She was awarded the Latin American Women in Chemistry Awards in the "Emerging Leader" category by the American Chemical Society and the Latin American Federation of Chemical Associations (FLAQ) in 2022.

Mónica's expertise is in the areas of fermentation processes, microbial biotransformation, enzyme production, industrial food waste valorization, extraction of bioactive compounds, and chemical characterization.

**Pedro Aguilar-Zarate, Ph.D.** is Professor at Tecnológico Nacional de México/Instituto Tecnológico de Ciudad Valles. He earned his Ph.D. degree in Food Science and Technology at Autonomous University of Coahuila, Mexico. His research focuses on the development of bioprocesses for the extraction and purification of secondary plant metabolites, the development of microbial fermentations in solid and liquid media to produce metabolites of industrial interest, and the study of protein-polyphenols interactions. Dr. Aguilar has published over 50 articles in indexed journals, over 30 book chapters, and nine articles in refereed journals. He has directed and co-directed 33 undergraduate students, 4 M.Sc. theses and 2 Ph.D. theses. He has made several presentations at scientific and academic events. He has been in charge of five research projects funded by the National Technological Institute of Mexico and collaborated on two projects funded by the sectoral funds SEP-CONACYT and CONAFOR-CONACYT. He is guest editor of *Frontiers in Food Science and Technology, Polymers, Plant Science Today, Foods* and the *International Journal of Research and Technological Innovation* (RIIT). This has earned him recognition as a member of the National System of Researchers level 2 (Mexico). His research and scientific career have been recognized by the Mexican Association of Food Science (2014), the Mexican Society of Biotechnology and Bioengineering (2015), the Mexican Association of Food Engineering and Biotechnology (2020), and the Potosino Council of Science and Technology (2021 and 2024).

**A. K. Haghi** is a retired professor and has written, co-written, edited, or co-edited more than 1000 publications, including books, book chapters, and papers in refereed journals with over 4200 citations and h-index of 34, according to the Google Scholar database. Professor Haghi holds a B.Sc. in urban and environmental engineering from the University of North Carolina (USA) and holds two M.Sc. degrees, one in mechanical engineering from North Carolina State University (USA) and another one in applied mechanics, acoustics, and materials from the Université de Technologie de Compiègne (France). He was awarded a Ph.D. in engineering sciences at Université de Franche-Comté (France).

Professor Haghi's extensive educational background and supervisory roles underscore his expertise and contributions to the field of engineering sciences. He is appointed as Honorary Research Associate (HRA) at the University of Coimbra, Portugal. He is a regular reviewer of leading international journals.

# Introduction to Food Waste-to-Energy Conversion Technologies

## 1.1 Understanding Food Waste-to-Energy Conversion Technologies

One of the objectives of the United Nations 2030 Agenda is to ensure access to affordable, safe, sustainable and modern energy (Goal 7). It is estimated that 660 million people will continue to lack access to electricity and that the trend for 2030 will be to continue depending on non-renewable energies such as fuels and oil.

It is necessary to implement strategies to address climate change and seek better, cleaner and more sustainable alternatives. According to the UN, energy continues to be the main cause of climate change, accounting for about 60% of global greenhouse gas emissions (United Nations Environment Programme, 2024).

To ensure that energy demand is met, it is vitally important to implement clean energy sources, modify ways of capturing energy, expand infrastructure and improve technology to supply clean energy around the world.

The circular economy has emerged as a model of production and consumption that seeks to reuse, renew and recycle materials and products as many times as possible, in such a way that they are used to the maximum, giving them added value. This avoids waste and reduces the generation of waste in production chains. The circular economy seeks to generate products efficiently and sustainably, reducing the demand for raw materials (Kalmykova et al., 2018; Velenturf & Purnell, 2021) (Fig. 1.1).

Waste is a serious global problem that brings with it various socio-economic problems. The search for strategies for the valorization of waste has been a task that has gained great interest not only in the scientific world but also at the industrial level to mitigate environmental change (Applied Energy, 2020).

Human activities, particularly industrial ones, have been based on the use of fossil fuels, which have been the energy basis for human development in recent decades. Today

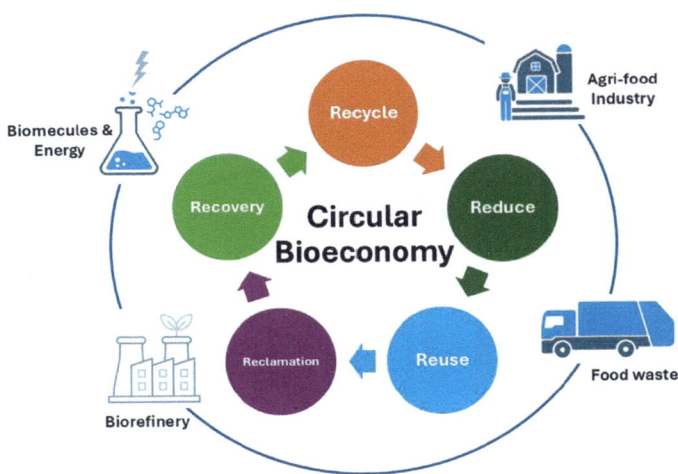

**Fig. 1.1** Circular bioeconomy as strategy to energy recovery in agri-food industry

the energy system operates on the basis of fossil fuels, 80% of global energy is obtained from fossil fuels (IRP, 2023).

Conventional forms of production must transition to more sustainable production models with a vision of social, economic, cultural and environmental responsibility.

Bioeconomy is considered necessary to solve global challenges (Kalmykova et al., 2018), covering the needs of human societies by making efficient use of natural resources, always maintaining balance in ecosystems.

One of the key factors to achieve this transition is the transformation and adaptation of sustainable value chains related to food (Arias et al., 2025). The agri-food industry is an important generator of waste; this waste has the characteristic of having a rich composition of materials that can be used as raw materials in various industries.

According to studies reported for the year 2019 more than 2000 million tons of municipal waste were generated (Kaza et al., 2018) of which just over 50% corresponded to food waste (Sarrion et al., 2023).

Food waste is defined as waste generated during the harvesting or harvesting stages up to the final finished producto (Food Waste Index Report, 2021; Jin et al., 2021). Depending on the type of product, there will be different processing steps and the type(s) of waste generated. Generally, each processing step releases waste. The agri-food industry is the main generator of waste; it is estimated that this industrial sector can release up to 40% of its waste (Food Waste Index Report, 2021; Wong et al., 2009).

It is estimated that approximately one-third of the food produced in the world is wasted through the production-circulation-consumption chain (Jin et al., 2021). According to FAO data, it is estimated that about 8.6% of cereals and pulses, fruits and vegetables and

## 1.1 Understanding Food Waste-to-Energy Conversion Technologies

21.6%, meat and animal products 11.9%, roots, tubers and oilseed crops 25.3% are lost and wasted annually (FAO, 2019).

By 2019, about 931 million tons of food waste were generated (Food Waste Index Report, 2021). In the United States alone, it is estimated that 30–40% of the food supply is considered food waste which corresponds to approximately 133 billion pounds which corresponded to 161 billion food in 2010 (Chakraborty et al., 2021). It is estimated that developing countries may discard up to 630 million tons of agri-food waste annually. Of the food waste generated globally, it is estimated that 61% comes from households, 26% from food services and 13% from retail (Global Economic Forum, 2021).

These materials are generally discarded and disposed of in the open air for the natural decomposition process to occur. However, this represents a major problem because without proper treatment of these wastes, one ton of material can generate up to 4.5 tons of $CO_2$ (United Nations, 2020). Incineration is another alternative to waste disposal but also represents disadvantages due to the water-rich composition of agro-industrial waste (Jin et al., 2021).

These wastes generate problems such as (1) correct disposal after processing, (2) emission of greenhouse gases as a result of the decomposition processes they undergo or burning, (3) contamination of aquiferous mantles due to leachate that leaks as a result of the decomposition of these residues (Borrello et al., 2017; Kumar et al., 2021).

Food losses are due to several reasons, among which the following stand out (Fig. 1.2):

(1) Inadequate management of farming techniques.
(2) Poor transportation and storage.
(3) Deficient infrastructure and production systems.
(4) Consumer behavior.
(5) Policies and regulations.

**Fig. 1.2** Main reasons for food losses

## 1.2 Rationale for Food-Waste-Based Energy

The linear economy is a model that presents problems such as the deterioration or depletion of natural systems (Attard et al., 2020), which also translates into economic losses due to the mass production of products that are discarded at the end of their useful life. And although the linear economy keeps the costs of services low, it is a system that overexploits natural resources and puts supplies at risk.

According to the International Resource Panel, the increase of natural resources used in the world has grown at an alarming rate, in just the last 50 years the depletion of resources has tripled, reaching a demand of 106.6 billion tons per year. This demand continues to grow at an accelerated rate of 2.3% per year, and at the current rate, the demand for natural resources is expected to double by 2050 (Global Resources Outlook 2024 Report).

The take-make-consume-discard consumption pattern contributes to the triple planetary crisis: (1) the climate change crisis, (2) the nature crisis and biodiversity loss and (3) the pollution and waste crisis (Hartley et al., 2020; Potočnik & Teixeira, 2023) (Fig. 1.3). Exacerbated consumption will exert catastrophic consequences on ecosystems in short periods of time.

The decrease in primary raw materials in convergence with the increase in waste generated and a culture of accelerated consumerism and discarding are important global problems that lead to a deterioration of the environment and therefore to economic and social systems. It is important to understand that sustainability is not only important for the conservation of natural resources but also for the conservation of the economy, societies and lifestyles that are currently in place around the world (ILO, 2015).

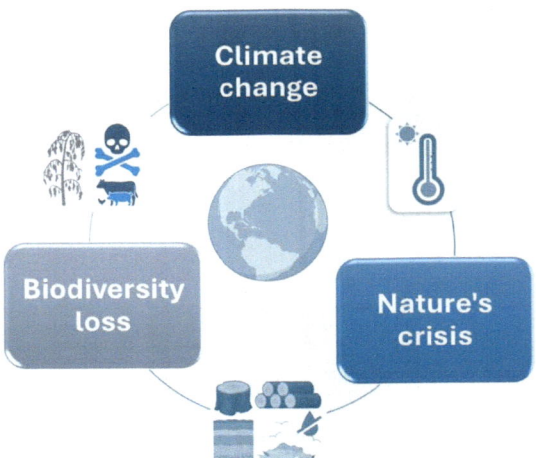

**Fig. 1.3** Triple global crisis

## 1.2 Rationale for Food-Waste-Based Energy

The Global Resources Outlook 2024 Report reveals that fossil fuels are the most traded commodity even though their extraction has decreased. It is not possible to maintain economic models based on fossil fuels because they are becoming increasingly scarce (Sharma et al., 2020) and their cost is high because they are becoming increasingly difficult to access due to overexploitation of these resources.

Changes in consumption/production patterns must be modified, creating an awareness of the use of natural resources. Recycling is not enough to alleviate the huge problem of accelerated consumption/production.

It is necessary to use natural resources in a responsible and forward-looking manner. Efforts must be made to use natural resources more profitably and for as long as possible. The energy transition should aim to reduce greenhouse gases and to achieve circularity in energy systems (IRP, 2023).

The bioeconomy has emerged with the objective of making processes more efficient to meet product and energy demands from sustainable processes (Mancini & Raggi, 2021; Sharma et al., 2020) and with the use of renewable raw materials that can be recyclable and biodegradable under an environmentally friendly approach.

Although there is a great academic development in the development of technologies that promote the circular economy, the appropriation of these technologies has been slow. It is estimated that globally only 9% is circular economy; in regions such as Europe the percentage rises to 12% and China with only 2%) (Hartley et al., 2020).

Among the strategies for reducing resource use are the food system and the energy system. In the energy system, it is recommended to decarbonize electricity, increase energy efficiency and increase the use of removable energy sources, particularly opting for the lowest cost ones (IRP, 2020). In the case of the food system, one of the recommendations is the reduction of losses and waste.

If the problem of the generation of a large amount of food waste and the need to diversify and generate clean energy are approached together, the idea of using food waste to obtain clean energy arises.

As previously mentioned, the food industry has a high potential for the energetic use of waste generated during the various production systems. Since of the total food supply (food supply) 40% becomes a loss or waste (Kaza et al., 2018). Food waste occurs at all stages of processing, from harvest and post-harvest, post-harvest and the various stages of food processing. Food waste should be avoided at all costs, and if it is generated, efficient strategies for its use should be sought.

Waste-to-energy is a new concept that seeks to manage waste with a sustainable vision (Klinhoffer & Castaldi, 2013). The circular bioeconomy seeks to make maximum use of resources by using the waste generated throughout the production processes in raw materials for the generation of other products (Velenturf & Purnell, 2021). This is why it is extremely important that, at the industrial level, the redesign of processes should be aimed at ensuring that the waste generated can give rise to new production lines in order to try to reduce the waste generated as much as possible. In other words, the generation

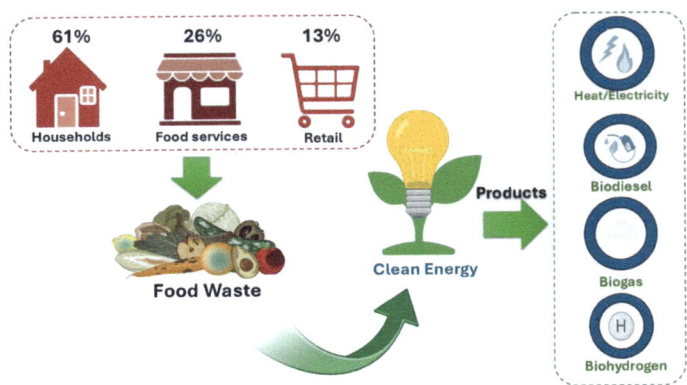

**Fig. 1.4** Diversification of clean energy products from food waste

of value chains from the waste of certain production processes is essential to generate a real bioeconomy (Gregg et al., 2020).

The conversion of food waste into energy is an effective strategy to address the excess of discarded organic materials in various production areas and to generate clean energy (Fig. 1.4).

The large quantities of organic biomass generated can give rise to the recovery/production of (1) chemicals and (2) energy. Biomasses have been mainly exploited for heat and power generation.

One of the major challenges of converting agro-industrial wastes into fuels and energy has been the problems related to the complexity and the physical and chemical composition of the various waste materials.

This complexity leads to the development of operational strategies, some of which are complex (depending on the type of waste and the quantity of derivatives to be obtained from the waste material). This complexity also entails costs in the recovery of the products of interest also associated with low conversion (Attard et al., 2020). These challenges have led to the development of numerous research projects to improve the efficiency of agro-industrial waste conversion processes into biocompostables, particularly under biorefinery concepts.

One of the most explored areas for converting food wastes into energy is anaerobic fermentation, although the production of biohydrogen, biomethane, liquid biofuels, electricity and commodity chemicals has also been explored (Sharma et al., 2020). These strategies have proven to be more efficient than conventional forms of waste management.

Although there are many benefits of using food waste for clean energy, it is necessary to consider the environmental impacts of these new processes (Gregg et al., 2020). The environmental impacts could vary depending on the technology used, the study of these impacts should be evaluated.

Supply–demand plays an important role in promoting new energy technologies. It is unquestionably necessary to migrate to clean energy, the appropriation of these new technologies will be marked by the supply and demand for them together with public policies that encourage their appropriation.

## 1.3   Environmental Benefits and Challenges

With the economic growth of some areas and the growing demand for resources, coupled with the fact that energy markets are fragile for various reasons, it is of great importance to work on an energy security plan, always seeking more efficient and less polluting energy systems (IRP, 2023).

Raising awareness of the effects of climate change in all social strata is of great importance; the economic development of today's societies cannot ignore respect for ecosystems. It is necessary to outline clear strategies to achieve energy security hand in hand with climate action. These actions must involve energy transition (van Summeren et al., 2020), moving away from fossil fuels and from non-renewable energy sources to renewable and increasingly cleaner technologies. More efficient and cleaner energy systems will make it possible to establish energy security.

In order to achieve energy security, broader approaches must be considered, going beyond the use of non-fossil fuels. The creation of value chains is essential to achieve sustainability and the promotion of the resilience of these productive chains is indispensable.

Energy security and climate action are inextricably linked; one cannot be achieved without the other. Therefore, the energy transition must be achieved primarily through the establishment of public policies at all levels of government around the world; these policies should encourage the use of clean energy from sustainable and sustainable supply chains.

The scientific community has made significant progress in the search for new technologies that seek to use non-traditional energy sources. These scientific and technological advances have triggered the development of efficient industries. It was reported that during the year 2023, the use of renewable energies will increase by more than 560 gigawatts (IEA, 2024).

Investment in clean energy is around two trillion dollars per year (double that of traditional energy sources). These data reveal the interest in achieving an energy transition. Clean energy sources are expected to generate half of all electricity by 2030 (IEA, 2024).

On the other hand, the high consumption of resources and their increase also predicts that the amount of industrial waste will increase, including food waste. It should be considered that although food waste can constitute a source for clean energy generation. Materials suitable for waste-to-energy conversion must be rich in lignocellulosic compounds. For this reason, the study of more food waste should be extended.

Waste-to-energy technologies should be increasingly efficient, robust, cost-effective and environmentally friendly (Applied Energy, 2020). Process optimization is a necessity in these conversion processes.

Another major challenge is the implementation of clean energy production processes in developing countries, where greater investment in infrastructure is required to make the new technologies cheaper. The valorization of food waste will undoubtedly help mitigate environmental pollution and take important steps towards sustainability.

## 1.4 Scope and Structure of the Book

This book explores energy recovery from food waste as a sustainable alternative for clean energy production.

Chapter one provides an introduction to food waste-to-energy techniques, an understanding of waste-to-energy conversion technologies, a rationale for the importance of producing energy from food waste, and the environmental benefits and challenges of these new technologies.

The characteristics of these waste materials to be considered as raw material for energy recovery will be addressed. The factors involved in the conversion of waste into energy will be explored in Chap. 2.

Chapter 3 explores case studies such as energy generation via combustion and its integration into the circular bioeconomy. It also explores the production of biodiesel, gasoline, and biofuels, as well as the existing regulations on the use of this type of energy.

In Chap. 4, you will find the production of biogas from food waste. This through anaerobic digestion, the most recent technologies for the production of fuels or clean energies are addressed. Information on the design and configuration of various types of reactors for biogas production will be found. Finally, you will find uses of the biogas produced as well as trends in biogas production.

## References

Applied Energy. (2020). Energy and resource recovery through integrated sustainable waste management. *261*, 114372.

Arias, A., Feijoo, G., Moreira, M. T., Tukker, A., & Cucurachi, S. (2025). Advancing waste valorization and end-of-life strategies in the bioeconomy through multi-criteria approaches and the safe and sustainable by design framework. *Renewable and Sustainable Energy Review, 207*, 114907.

Attard, T. M., Clark, J. H., & McElroy, C. R. (2020). Recent developments in key biorefinery areas. *Current Opinion in Green and Sustainable Chemistry, 21*, 64–74. https://doi.org/10.1016/j.cogsc.2019.12.002

Borrello, M., Caracciolo, F., Lombardi, A., Pascucci, S., & Cembalo, L. (2017). Consumers' perspective on circular economy strategy for reducing food waste. *Sustainability, 9*(1), 141. https://doi.org/10.3390/su9010141

# References

Chakraborty, D., Althuri, A., Chatterjee, S., & Mohan, V. (2021). Bioconversion technologies: Hydrolytic enzyme treatment of food waste. In *Executive summary—World energy outlook 2024—Analysis—IEA*.

FAO. (2019). *The state of Food and Agriculture 2019. Moving forward on food loss and waste reduction*. Rome. Licence: CC BY-NC-SA 3.0 IGO. ISBN: 978-92-5-131789-1. https://www.fao.org/interactive/state-of-food-agriculture/2019/en/

Global Economic Forum. (2021). *The world's food waste problem is bigger than we thought-here's what we can do about it*. Entrance. March, 2021. https://www.weforum.org/stories/2021/03/global-food-waste-solutions/. January 18th 2021.

Gregg, J. S., Jürgens, J., Happel, M. K., Strom-Andersen, N., Tanner, A. N., Bolwing, S., & Klitkou, A. (2020). Valorization of bio-residuals in the food and forestry sectors in support of a circular bioeconomy: A review. *Journal of Cleaner Production, 267*, 122093.

Hartley, L., van Santen, R., & Kirchherr, J. (2020). Policies for transitioning towards a circular ecnomoy: Expectations from the European Union (EU). *Resources, Conservation & Recycling, 155*, 104634.

IEA. (2024). *World energy outlook 2024*. IEA, Paris https://www.iea.org/reports/world-energy-outlook-2024. Licence: CC BY 4.0 (report); CC BY NC SA 4.0 (Annex A). https://www.iea.org/reports/world-energy-outlook-2024/executive-summary

International Labour Organization. (2015). *Guidelines for a just transition towards environmentally sustainable economies and societies for all*. https://www.ilo.org/global/topics/green-jobs/publications/WCMS_432859/lang--en/index.htm

IRP. (2020). Resource efficiency and climate change: Material efficiency strategies for a low-carbon future. In E. Hertwich, R. Lifset, S. Pauliuk, N. Heeren (Eds.), *A report of the international resource panel*. United Nations Environment Programme, Nairobi, Kenya.

IRP. (2023). Enabling the energy transition: Mitigating growth in material and energy needs, and building a sustainable mining sector. In J. Potočnik, & I. Teixeira (Eds.), *An opinion piece of the international resource panel co-chairs*. https://www.resourcepanel.org/reports/enabling-energy-transition

Jin, C., Sun, S., Yang, D., Sheng, W., Ma, Y., He, W., & Li, G. (2021). Anaerobic digestion: An alternative resource treatment option for food waste in China. *Science of the Total Environment, 779*, 146397.

Kalmykova, Y., Sadagopan, M., & Rosado, L. (2018). Circular economy—From review of theories and practices to development of implementation tolos. *Resources, Conservation & Recycling, 135*, 190–201.

Kaza, S., Yao, L. C., Bhada-Tata, P., & Van Woerden, F. (2018). What a waste 2.0: A global snapshot of solid waste management to 2050. Urban Development; © World Bank, Washington, DC. http://hdl.handle.net/10986/30317. License: CC BY 3.0 IGO.

Klinhoffer, N. B. & Castaldi, M. J. (2013). Waste to energy conversion technology. ISBN: 978-0-85709-011-9. Woodhead Publishing

Kumar, M., Dutta, S., You, S., Luo, G., Zhang, S., Show, P. L., Sawarkar, A. D., Singh, L., & Tsang, D. C. W. (2021). A critical review on biochar for enhancing biogas production from anaerobic digestion of food waste and sludge. *Journal of Cleaner Production, 305*, 127143.

Mancini, E., & Raggi, A. (2021). A review of circularity and sustainability in anaerobic digestion processes. *Journal of Environmental Management, 291*, 112695.

Potočnik, J., & Teixeira, I. (2023). An opinion piece of the international resource panel co chairs.

Sarrion, A., Medina-Martos, E., Iribarren, D., Diaz, E., Mohedano, A. F., & Dufour, J. (2023). Life cycle assessment of a novel strategy based on hydrothermal carbonization for nutrient and energy recovery from food waste. *Science of the Total Environment, 878*, 163104.

Sharma, S., Basu, S., Shetti, N. P., & Aminabhavi, T. M. (2020). Waste-to-energy nexus for circular economy and environmental protection: Recent trends in hydrogen energy. *Science of the Total Environmental, 713*, 136633.

United Nations. (2020). Food loss, waste account for 8 per cent of all greenhouse-gas emissions, says deputy secretary-general, marking inaugural international awareness day. 2020. DSG/SM/1465 29 September 2020 https://press.un.org/en/2020/dsgsm1465.doc.htm.

United Nations Environment Programme. (2021). *Food waste index report 2021*. Nairobi. https://www.unep.org/resources/report/unep-food-waste-index-report-2021

United Nations Environment Programme. (2024). *Global resources outlook 2024: Bend the trend—pathways to a liveable planet as resource use spikes.* International Resource Panel, Nairobi. https://wedocs.unep.org/20.500.11822/44901

van Summeren, L. F. M., Wieczorek, A. J., Bombaerts, G. J. T., & Verbong, G. P. J. (2020). Community energy meets Smart grids: Reviewing goals, structure, and roles in Virtual Power Plants in Ireland, Belgium and the Netherlands. *Energy Research & Social Science, 63*, 101415.

Velenturf, A. P. M., & Purnell, P. (2021). Principles for a sustainable ciruclar economy. *Sustainable Production and Consumption, 27*, 1437–1457. https://doi.org/10.1016/j.spc.2021.02.018

Wong, J. C., Fung, S. O., & Selvam, A. (2009). Coal fly ash and lime addition enhances the rate and efficiency of decomposition of food waste during composting. *Bioresource Technology, 100*(13), 3324–3331. ISSN 0960-8524. https://doi.org/10.1016/j.biortech.2009.01.063

# Sustainable Approaches and Characteristics 2

## 2.1 Types of Food Waste-to-Energy Production

According to the Food Waste Index Report of the United Nations Environment Program, food waste is estimated at 931 million tons annually. Of which about 570 million tons are generated in households (Food Waste Index Report, 2021).

Food wastage is around 74 kg per capita and there are no significant differences between the degree of economic development of the countries studied (Food Waste Index Report, 2021). This study also reveals that these estimates may be underestimated since many of the countries studied do not have sufficient information. Data are insufficient as there is a knowledge gap regarding the standardization of data collection methodologies (Sjölund et al., 2025).

It is estimated that middle- and high-income countries show no difference in terms of household waste (76–79%) (World Resources Report, 2019). Regarding the waste generated by food services and retail, the percentages range from 26 to 13% respectively. What is a fact is that food waste worldwide is a problem that requires the search for innovative solutions that promote sustainable management of these materials. This report is the first of its kind, which makes it difficult to know accurate worldwide statistics.

Food waste is defined as food and inedible parts (bones, rinds, pits) of food that may be discarded from the food supply chain (Cisse et al., 2025).

These wastes can be generated throughout the entire production chain, from processing to transportation, points of sale, food services, and from processing, transportation, points of sale, food services or households. Food waste can be solid or liquid, raw or cooked food substance that is discarded along the food chain.

These wastes usually end up in landfills, as garbage, and at best in landfills or combustion. It is estimated that if converted into calories, this waste would be equivalent to

24% of the world's wasted food supplyfood supply that is wasted. These losses are devastating, not only because of the loss of the feed itself but also because of the loss of the natural sources that were necessary to grow the raw material, the investment during processing, but also because of the loss of the natural sources that were necessary to grow the raw material, the investment during processing and the transportation costs. This food loss consumes about a quarter of all water used in agriculture annually (World Resources Report, 2019). In addition to the fact that to produce this food that was not consumed required about 1.4 billion hectares of land used to produce food that is not consumed (Moonsamy et al., 2024).

This represents a major problem with several aspects, such as world hunger or the water stress in which many regions of the world find themselves. If food loss were a country, it would be the third country with the highest GreenHouses Gases emissions with about 8% (Eves et al., 2025). According to FAO (2011), 56% of food losses occur in developed regions, with industrialized Asia leading the list with 28% worldwide.

## 2.2 Composition of Food Waste

Food waste is a source of organic materials composed mainly of carbohydrates, cellulose, hemicellulose, lipids and proteins (Moonsamy et al., 2024) (Fig. 2.1). Depending on the type of food waste, the content of each of these fractions may vary. The effect of these compositional variations on food waste conversion treatments has been demonstrated (Gu et al., 2024). The greater the heterogeneity of the waste, the more complex the design of the strategy to recover the materials will have to be.

Food waste is a rich source of carbon with high carbon-to-nitrogen and carbon-to-phosphorus ratios; these characteristics make it an excellent raw material for fermentation and high energy recovery (Table 2.1). They also have a rich component composition that makes their extraction desirable. These residual materials are highly biodegradable. They possess a high moisture content (70–80%) and a pH in the range of 4–6 (Ye et al., 2024).

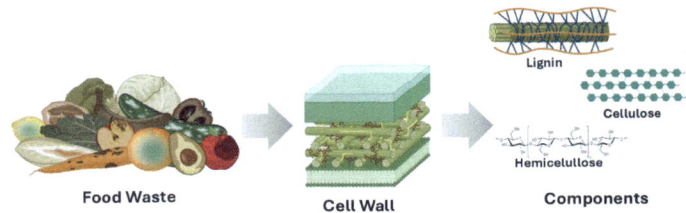

**Fig. 2.1** Main components of food waste

## 2.2 Composition of Food Waste

**Table 2.1** Main composition of some food waste Talavera-Caro et al., (2020) and Rojas et al., (2025) with modifications

| Food waste | Cellulose % | Lignin % | Hemicellulose % | C % | N % | C:N |
|---|---|---|---|---|---|---|
| Rice straw | 35–44 | 12–13 | 27–34 | 39.7 | 0.9 | 47–67 |
| Sugarcane bagasse | 40–45 | 25–30 | 20–24 | 46.08 | 0.74 | 118:150 |
| Corn stover | 40 | 14–17 | 25–31 | 43.2 | 0.8 | 50–63 |
| Corn cob | 45 | 15 | 25 | 41.26 | 0.45 | 123 |
| Rice husk | 48 | 25 | 32.0 | 45.8 | 0.3 | |
| Cocoa pod | 18–23 | 32.5 | 15 | 43.87 | 0.17 | |
| Cocoa bean shells | 24–35 | 14.5–26.5 | 9–11 | 50.3 | 3.5 | |
| Coffee husk | 19–26 | 18–30 | 24–45 | 46.8 | 0.6 | |
| Spent coffee grounds | 61.5 | 28 | | 54.9 | 3.5 | |

According to Kassim et al. (2025), elemental analysis of FW should be carried out. Determining the carbon, nitrogen, oxygen, and hydrogen content, as well as the ash content, is essential to knowing the degree of conversion to bioenergy.

There are differences in compositional content depending on the type of feed residue. For example, animal-based food waste is rich in protein, lipid, and ash content. While plant-based residues are rich in higher heating value, high in carbohydrates, starch and lignocellulosic material. Within the plant-based residues, a distinction can be made between lignin-rich residues and starch-poor residues (Nazibudin et al., 2025). FW are characterized by their high water content (Nikiema et al., 2022).

Understanding the composition of food waste is a vital step towards choosing the best way to handle the waste material. Not all food waste has the right characteristics to be used as raw material for clean energy conversion.

Materials richer in cellulose and hemicellulose have been shown to have the greatest potential for clean energy production. In order to recover hydrogen with higher yields, it is necessary to use materials with a high carbohydrate and cellulose content (Gu et al., 2024).

### 2.2.1 Pretreatments Used on Food Wastes

As mentioned above, a large amount of food waste is rich in lignocellulosic materials that require pretreatment leading to delignification prior to fermentation processes in order to be used. Pretreatments are the basis for depolymerization of lignocellulosic material present in food waste to monomers such as glucose so that it can be used as a substrate in fermentation processes. These pretreatments reduce the crystallinity of the lignin, which means that less organized structures can be obtained, allowing the action of hydrolytic

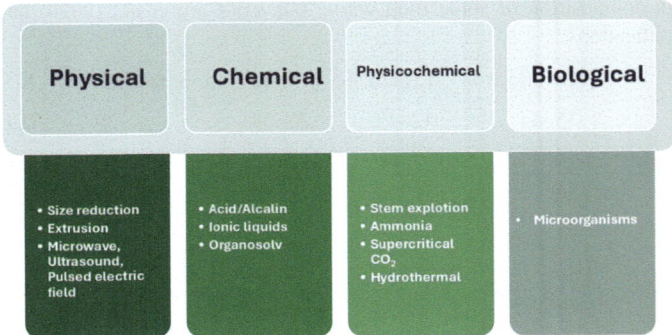

**Fig. 2.2** Pretreatments for lignocellulosic materials

enzymes either added to the reactors or by the action of the enzymes released when the microorganisms grow.

There are several types of pretreatments that can be divided into four categories: Physical, Chemical, Physicochemical and Biological (Fig. 2.2).

The purpose of these pretreatments is to condition the raw materials through particle size reduction in order to increase the contact surface of the FW. The pretreatments allow the modification of its rigid structure through the variation of the pH with the addition of acids, bases, or a series of organic solvents that even allow the separation of the fractions of interest. Other methodologies include radiation with ultrasound or microwaves that allow exposure of the cellulose to make it more available for degradation. Technologies such as stem explosion, supercritical fluids, hydrothermal treatments allow a more aggressive degradation of the lignin to obtain monomers by breaking the fibers that give the FW structure (Aguirre-Fierro et al., 2020).

## 2.3 Food Waste to Energy Research and Development

The production of bioenergy from FW is possible, a reality that is becoming increasingly attractive due to the environmental and economic benefits it represents. This approach provides a solution to the problems of pollution and FW management and disposal and is capable of producing bioenergy (Kassim et al., 2025).

In order to valorize food wastes, it is necessary to consider the type of waste generated in order to categorize the type of conversion to which it could be subjected. According to Elbersen et al. (2015) the type of FW should be classified into those for: (1) thermal conversion technologies, (2) anaerobic digestion and (3) biochemical conversion.

In general terms, the conversion technologies that could be applied to WF to obtain energy could be divided into biological and thermochemical (Fig. 2.3). Through these

## 2.4 Thermochemical Conversion

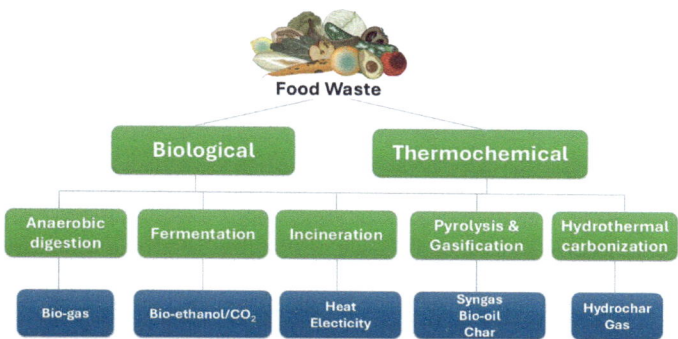

**Fig. 2.3** Food waste to energy conversion technologies

conversions, several types of clean energy can be obtained, among which the following stand out: heat/energy, Biogas, Biodiesel, Biohydrogen, Anaerobic digestion, bioethanol, syngas, hydrochar, hydrothermal carbonization.

If the composition of the FW is rich in lignocellulosic materials there are two types of conversion technologies generally used for the valorization of these materials (Hassan et al., 2019).

(1) Biochemical conversion of lignocellulose. This involves the hydrolysis of the materials to then lead or not to a fermentative process.
(2) Thermochemical conversion. This involves direct combustion, pyrolysis, gasification or torrefaction.

## 2.4 Thermochemical Conversion

Thermochemical conversion is another option to FW management. In this technology, combustion, gasification, liquefaction, pyrolysis, torrefaction can be carried out. The most advanced technologies are fast pyrolysis and HTPs (Rojas et al., 2025).

Thermochemical conversion involves high costs due to the high pressures and temperatures involved in the process. Gasification and pyrolysis are the most commonly used conversions, in these processes the chemical bonds are broken to achieve the transformation into biochar, bio-oil and syngas, which are the basis for the formation of biofuels such as biodiesel, biohydrogen, bioethanol and biomethane (Rojas et al., 2025).

In this type of process, the moisture content is important the moisture content is important, since the contact between water and biomass during the process can be affected (Rojas et al., 2025).

One of the limitations of this type of conversion are the pretreatments to which the FW must be subjected. Another limitation that must be considered is the prevention of gases such as $NO_X$, $SO_X$ and particulate matter and heavy metals (Sarrion et al., 2023).

## 2.5 Anaerobic Digestion

Anaerobic digestion is one of the most mature, effective, and widely used technologies for treating food waste. This technology allows the conversion of a wide variety of food waste into two important products. One of them is a digestate rich in nutrients, a digestate rich in nutrients and the other in clean energy (Gu et al., 2024; Sarrion et al., 2023).

Gases such as methane, hydrogen, and carbon dioxide can be recovered from anaerobic digestion. Anaerobic digestion consists of two phases:

(1) Black fermentation. Production of volatile organic acids and hydrogen.
(2) Methanogenic. Methane production.

Power generation proves to be more cost-effective due to lower infrastructure and processing costs compared to other clean energy production technologies (Li, 2024).

The use of anaerobic digestion for gas production is a reality in Europe, where given the high waste rates it can be used for biogas production. This type of conversion technology has several disadvantages, the most important factor being time, since long fermentation periods are required to produce the gases. Among the parameters that must be carefully monitored are the accumulation of $NH_3$ during fermentations, pH changes, and the effects on the metabolism of microorganisms. Of the technologies applied to food waste conversion, anaerobic digestion has been shown to have less environmental impact than other technologies. It has the lowest global warming, acidification and eutrophication potential. Anaerobic digestion has a considerable reduction of $CO_2$ generation compared to composting (Jin et al., 2021).

## 2.6 Biochemical Conversion

The purpose of this conversion is to pretreat the materials, especially those rich in lignocellulose, to generate depolymerization and obtain polysaccharides that can be converted into biofuels or some other chemical product of interest. In this type of conversion it is essential to know the content of indicators such as lignin content, cellulose, hemicellulose and ash.

A wide variety of microorganisms (fungi, bacteria, yeasts, algae) and enzymes can be used in biochemical technology. If whole microbial cells are used to take advantage of FW it is called fermentation. During this process, the microorganisms take advantage of

## 2.6 Biochemical Conversion

the waste as fermentation substrates; this implies that their growth generates, through their metabolic capabilities, the synthesis and expression of enzymes that manage to hydrolyze the complex composition of the food waste. From these fermentations, enzyme extracts may or may not be recovered, concentrated and/or purified to have enzyme concentrates that can be applied to other substrates under standardized methodologies to maximize the potential use of these enzymes on a wide variety of substrates.

Biochemical methods such as fermentation and enzymatic hydrolysis are good technologies for converting food waste into energy (Rojas et al., 2025). The limitation is the high cost of the processes due to the use of enzymes, low specificity of the enzymes for complex substrates, which leads to low conversion yields (Rojas et al., 2025).

Bio-ethanol is usually obtained through this type of biochemical methods as a source of energy for gas and thermal engines (Sridhar et al., 2021). Production costs can be high due to the distillation required to recover the ethanol produced. Food wastes with high carbohydrate and cellulose content are usually the most suitable for better conversion yields.

In this technology, fermentation optimization must be considered for the design of an adequate FW management.

### 2.6.1 Biorefinery Concept

If biofuels are to be obtained from FW, the biorefinery is an excellent alternative. Biorefinery, which is defined as the sustainable transformation of biomolecules into a wide range of bioproducts and bioenergy, has been one of the bases for the development of more sustainable economic systems (Hassan et al., 2019). The production of biofuels from food waste under the biorefinery concept are classified as second generation fuels (Hota & Maiti, 2024).

If biofuels are to be obtained from food wastes, the conversion technologies to be applied are hydrolysis, fermentation, transesterification and enzymatic saccharification. These technologies can be used to obtain fuels such as biohydrogen, biodiesel, biobutanol and bioetanol (Ye et al., 2024).

### 2.6.2 Bioelectricity Generation

This technology is interesting and promising, consisting of using microorganisms capable of producing electricity from food waste (Kakkar et al., 2024). The technology is based on the extracellular transfer of electrons that are generated by the interaction of microbial metabolism with electrodes. Energy and other valuable products are recovered from this interaction (Akram et al., 2025). Several studies have addressed the generation of bioelectricity from food waste (Rojas-Flores et al., 2022).

## 2.7 Factors Affecting Food Waste-to-Energy Conversion

One of the most influential factors in the conversion processes is related to the variable composition of the FW, depending on the composition will be the degree of conversion. In order to standardize the conversion processes, it is necessary to characterize food wastes in detail. This is perhaps the greatest limitation, because the heterogeneity of the waste is great and varies by source, time of year, geographical region, etc. (Kassim et al., 2025).

## 2.8 Thermal Conversion System

For thermal conversion systems, there are challenges that determine the conversion quality of FW to energy. The first one is the ash content that could affect energy production if found in high concentrations. This parameter is important because when scaling up processes, the infrastructure could be limited by its design or even put the equipment at risk due to wear and tear.

There are four factors that determine the conversion quality of food waste to energy: (1) corrosion, (2) ash content, (3) ash melting and (4) nitrogen emissions.

### 2.8.1 Ash Melting

Ash agglomeration is another major problem as it causes equipment fouling. Equipment is usually designed to contain limited amounts of ash. This challenge can be overcome with ash melting indicator sensors.

### 2.8.2 Corrosion

Corrosion is determined by the chlorine content and the Cl:S ratio present in the food waste.

### 2.8.3 Nitrogen Content

Another parameter to consider is the nitrogen content, as there may be nitrogen emissions during conversion. Emissions of nitrogen oxides are a problem because they generate air pollution and eutrophication.

In addition, depending on the country where the technology is located, taxes could be paid, making the technology more expensive. The latter, especially if they are small-scale equipment (Lammens et al., 2016).

## 2.9 Biochemical Conversion

This process involves the pretreatment of lignocellulosic materials, from which the production of fuels and chemicals is derived. The most important parameters for this type of conversion are:

### 2.9.1 Lignin Content

A high lignin content will hinder biochemical conversion to cellulose and hemicellulose.

### 2.9.2 Cellulose and Hemicellulose Content

These components are of greatest interest for biochemical conversion, so it is preferred that they are in high concentrations in order to recover more sugars.

### 2.9.3 Ash Content

High ash content can increase costs in processing due to equipment damage or disposal/handling costs (Elbersen et al., 2015).

## 2.10 Anaerobic Digestion

In anaerobic digestion the most relevant factors are the limitations by the type of reactor used and specifically by the volume. The volume of the reactor will determine the conversion yield. The larger the reactor the higher the product yield.

The performance will be affected by the composition of the FW, so the selection of the type of residue should be an important factor to consider when designing processes with this technology. The selection of the pretreatment to which the feed residue will be subjected will affect the mass transfer and therefore the good development of anaerobic digestion. Another parameter to be taken care of in anaerobic digestion is the salt content of the waste, if the content of ions such as sodium, potassium, calcium and magnesium is high, they could inhibit anaerobic digestion (Pham et al., 2015).

The feed residue must be known because during anaerobic digestion, in addition to the production of biogas, a "digestate" is generated which, depending on its composition, could be used as fertilizer (Sarrion et al., 2023).

Thermochemical methodology, anaerobic digestion, fermentation and bioelectrochemical systems have been considered as the most promising alternatives for energy conversion either through microogranisms or chemical conversion.

## References

Aguirre-Fierro, A., Pino, M. S., Zanuso, E., Londoño-Hernández, L., Nájera, A., Torres, A. Y., Aguilar, C. N., Rodríguez-Jasso, R. M., Robledo-Olivo, A., & Ruiz, H. A. (2020). Biochemical and thermochemical platforms for bioproducts and biofuels in terms of biorefinery. In *Advances in food bioproducts and bioprocessing technolgies* (1st ed., pp. 145–192). CRC Press. ISBN: 9780429331817.

Akram, F., Salamat, D., Fatima, T., & Shabbir, I.-u-H. (2025). Recent redns, perceptions and applications of microbial bioelectrochemcial systems: An innovative technology. *Journal of Electroanalytical Chemistry, 979*(15), 118933. https://doi.org/10.1016/j.jelechem.2025.118933

Cisse, R. S., Aguilar, F. S., Ezenagu, S. B., Cisse, M. S., & Niang, A. (2025). Characterization of food waste in Grizzly Dining Hall at Georgia Gwinnett College: A critical step toward sustainable food waste management. *Heliyon, 11*, e41750.

Elbersen, W., Bakker, R., Harmse, P., Vis, M., & Alakangas, E., (2015). A selection method to match biomass types with the best conversion technologies. S2BIOM Deliverable D2.2. S2Biom Project Grant Agreement n° 608622. https://www.s2biom.eu/images/Publications/D2.2_S2Biom_Selection_methodology_for_matching_Final2.pdf. https://www.s2biom.eu/en/publications-reports/s2biom.html

Eves, A., Kim, B., Hodgkins, C., Raats, M., & Timotijevic, L. (2025). Is it food or is it waste? Determinants of decisions to throw food away. *Sustainable Production and Consumption, 54*, 43–51.

FAO. (2011). *The state of food and agriculuture: Women in agriculture-closing the gender gap for development*. FAO, Rome. https://research.wri.org/sites/default/files/2019-07/WRR_Food_Full_Report_0.pdf

Gu, S., Xing, H., Zhang, L., Wang, R., Kuang, R., & Li, Y. (2024). Effects of food wastes ased on different components on digestibility and energy recovery in hydrogen and methane co-production. *Heliyon, 10*, e25421. https://doi.org/10.1016/j.heliyon.2024.e25421

Hassan, S. S., Williams, G. A., & Jaiswal, A. K. (2019). Moving towards the second generation of lignocellulosic biorefineries in the EU: Drivers, challenges, and opportunities. *Renewable and Sustainable Energy Reviews, 101*, 590–599.

Hota, P., & Maiti, D. K. (2024). Chemistry for catalytic conversion of biomass/waste into green fuels. In *Sustainable green catalytic processes*. Wiley. Online ISBN: 9781394212767. Print ISBN: 9781394212552. https://doi.org/10.1002/9781394212767.ch15

Jin, C., Sun, S., Yang, D., Sheng, W., Ma, Y., He, W., & Li, G. (2021). Anaerobic digestion: An alternative resource treatment option for food waste in China. *Science of the Total Environment, 779*, 146397.

Kakkar, S., Dharavat, N., & Sudabattula, S. K. (2024). Transforming food waste into energy: A comprehensive review. *Results in Engineering, 24*, 103376.

Kassim, F. O., Sohail, M., Somorin, T., Blanch, G., Yaman, R., & Afolabi, O. O. D. (2025). Optimised mixed agri-food waste simulant for enhanced bioenergy production via hydrothermal carbonization and supercritical plant modelling. *Energy Reports, 13*, 184–195.

# References

Lammens, T., Vis, M., van den Berg, D., de Groot, H., Vanmeulebrouk, B., Staristsky, I., Annevelink, B., Elbersen, W., & Elbersen, B. (2016). *S2Biom deliverable D4.5. Bio2Match: A tool for matching biomass and conversion technologies.* S2Biom Project Grand Agreement n° 608622. https://www.s2biom.eu/en/publications-reports/s2biom.html. https://s2biom.wenr.wur.nl/doc/S2Biom%20D4.5%20-%20Tool%20for%20Matching%20Biomass%20and%20Conversion%20Technologies.pdf

Li, R. (2024). Techno-economic and environmental characterization of municipal food waste-to-energy biorefineries: Integrating pathway with compositional dynamics. *Renewable Energy, 223*, 120038. https://doi.org/10.1016/j.renene.2024.120038

Moonsamy, T. A., Rajauria, G., Priyadarshini, A., & Jansen, M. A. K. (2024). Food waste: Analysis of the complex and variable composition of a promising feedstock for valorisation. *Food and Bioproducts Processing, 14*, 31–42.

Nazibudin, N. A., Sabri, S. A. M., & Manaf, L. A. (2025). Aligning with sustaible development goals (SDG) 12: A systemic review of food waste generation in Malaysia. *Cleaner Waste Systems, 10*, 100205.

Nikiema, J., Asamoah, B., Egblewogbe, M. N. Y. H., Akomea-Agyin, J., Cofie, O. O., Hughes, A. F., Gebreyesus, G., Asiedu, K. Z., & Njenga, M. (2022). Impact of material composition and food waste decomposition on charcteristics of fuel briquettes. *Resources, Conservation & Recycling Advances, 15*, 200095.

Pham, T. P. T., Kaushik, R., Parshetti, G. K., Mahmood, R., & Balasubramanian, R. (2015). Food waste-to-energy conversion technologies: Current status and future directions. *Waste Management, 38*, 399–408. https://doi.org/10.1016/j.wasman.2014.12.004

Rojas, M., Manrique, R., Hornung, U., Funke, A., Mullen, C. A., Chejne, F., & Maya, J. C. (2025). Advances and challenges on hydrothermal processes for biomass conversion: Feedstock flexibility, products, and modeling approaches. *Biomass and Bioenergy, 194*, 107621.

Rojas-Flores, S., De La Cruz-Noriega, M., Nazario-Naveda, R., Benites, S. M., Delfín-Narciso, D., Rojas-Villacorta, W., & Romero, C. V. (2022). Bioelectricity through microbial fuel cells using avocado waste. *Energy Reports, 8*(9), 376–382. https://doi.org/10.1016/j.egyr.2022.06.100

Sarrion, A., Medina-Martos, E., Iribarren, D., Diaz, E., Mohedano, A. F., & Dufour, J. (2023). Life cycle assessment of a novel strategy base don hydrothermal carbonization for nutrient and energy recovery from food waste. *Science of the Total Environment, 878*, 163104.

Sjölund, A., Malefors, C., Svensson, E., von Brömssen, C., & Eriksson, M. (2025). Rethinking household food waste quantification: Increasing accuracy and reducing costs through automation. *Environmental Technology & Innovation, 37*, 103993.

Sridhar, A., Kapoor, A., Kumar, P. S., Ponnuchamy, M., Balasubramanian, S., & Prabhakar, S. (2021). Conversion of food waste to energy: A focus on sustainability and life cycle assessment. *Fuel, 302*, 121069.

Talavera-Caro, A. G. et al. (2020). Proteomics of lignocellulosic substrates bioconversion in anaerobic digesters to increase carbon recovery as methane. In Z. Zakaria, R. Boopathy, & J. Dib (Eds.), *Valorisation of agro-industrial residues—Volume I: Biological approaches. Applied environmental science and engineering for a sustainable future.* Springer, Cham. https://doi.org/10.1007/978-3-030-39137-9_4

United Nations Environment Programme. (2021). *Food waste index report 2021.* Nairobi.

Ye, Y., Guo, W., Ngo, H. H., Wei, W., Cheng, D., Bui, X. T., Hoang, N. B., & Zhang, H. (2024). Biofuel production for circular bioeconomy: Present scenario and future scope. *Science of the Total Environmental, 935*, 172863.

World Resources Report. (2019). Creating a sustainable Food Futre. A menú of Solutions to Feed Neary 10 Billion People by 2025. ISBN 978-1-56973-963-1. https://www.wri.org/research/creating-sustainable-food-future

# Current Conversion Technologies and Case Studies

## 3.1 Introduction

Agro-food waste is a valuable resource with significant energy recovery potential. The integration of advanced technologies and real-world applications, as illustrated in these case studies, is paving the way for a sustainable energy future. These methods not only reduce waste but also help meet energy demands, supporting both environmental and economic goals. If you want to dive deeper, the references provided in each section can offer detailed insights into these innovative practices.

Energy recovery from agro-food waste involves converting organic waste generated by agricultural and food industries into usable energy forms like biogas, biochar, or electricity through different methods.

Thermochemical methods use heat to break down organic waste into energy-rich products. For example, pyrolysis is a process where waste is heated in the absence of oxygen to produce biochar, syngas, and bio-oil. Slow pyrolysis of agro-food waste at around 300 °C has shown promising results in increasing energy recovery efficiency (Economou et al., 2024).

Biological methods rely on microorganisms or enzymes to break down organic matter. Anaerobic digestion uses bacteria to decompose waste in the absence of oxygen, producing biogas (a mixture of methane and carbon dioxide) that can be used for electricity and heat. This process is common for treating agricultural residues and food waste. Microbial Fuel Cells (MFCs) are innovative systems that use microbes to convert organic waste directly into electricity. While still developing, MFCs represent a sustainable solution to treating wastewater and generating energy simultaneously (Ramanaiah et al., 2023).

Biorefineries combine various processes to maximize waste utilization and resource recovery. These facilities can process agro-food waste to produce multiple products, such

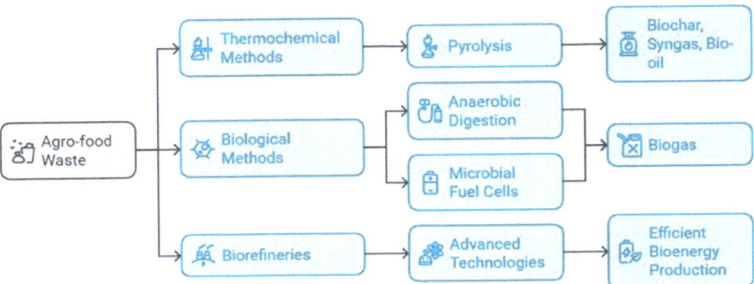

**Fig. 3.1** General overview for the transformation of agro-food waste into energy

as biofuels, biochar, and value-added chemicals like biofertilizers. Incorporating nanotechnology and biotechnology, biorefineries enhance the efficiency and range of usable outputs (El-Ramady et al., 2022).

Emerging technologies focus on using advanced processes to overcome limitations of traditional methods. Microbial conversion takes advantage of microorganisms that naturally metabolize waste, producing energy during the process. For instance, fruit waste and other organic residues can be treated with microbial systems for efficient bioenergy production (Borja & Fernández-Rodríguez, 2021). Technologies like gasification, which involves partial combustion of waste at high temperatures, yield syngas (a fuel gas mixture) that can be used for power generation.

The agro-food sector generates large quantities of wastewater laden with organic materials. This wastewater can be treated using Microbial Fuel Cells. These cells recover energy while cleaning the water, making them a dual-purpose technology. However, challenges such as scalability and cost remain significant (Cecconet et al., 2018) (Fig. 3.1; Table 3.1).

## 3.2 Energy Recovery from Food Waste by Combustion

Energy recovery through combustion involves burning food waste to generate heat, which can be converted into electricity or used directly for industrial processes. While it is a long-established method, modern advancements and integration with sustainable waste management practices have refined its applications.

Food waste typically contains high moisture content and organic material, which can be burned under controlled conditions in waste-to-energy (WTE) plants:

- High-Temperature Combustion: In this process, food waste is incinerated at temperatures above 850 °C. This ensures complete oxidation, reducing waste volume by up to 90%.

## 3.2 Energy Recovery from Food Waste by Combustion

**Table 3.1** Food-waste conversion technologies and case studies

| Technology | Description | Case study | Reference |
|---|---|---|---|
| Pyrolysis | Thermal decomposition of waste into biochar and syngas in the absence of oxygen | Slow pyrolysis of food waste in Portugal | Economou et al. (2024) |
| Gasification | Partial oxidation of waste to produce syngas | Energy recovery from agricultural residues | Berenguer et al. (2023) |
| Anaerobic digestion | Microbial decomposition of waste in oxygen-free environments, producing methane-rich biogas | Agricultural waste biogas plants in Spain | Borja and Fernández-Rodríguez (2021) |
| Microbial fuel cells (MFCs) | Electricity generation via microbial metabolic activities while treating wastewater | Agro-food wastewater treatment in Italy | Cecconet et al. (2018) |
| Fermentation | Conversion of sugars to ethanol or other biofuels through microbial action | Ethanol production from fruit waste | Ramanaiah et al. (2023) |
| Nanotechnology applications | Enhancing waste conversion processes, such as biofuel stability improvements | Nanotechnology in waste valorization | El-Ramady et al. (2022) |
| Enzymatic valorization | Use of enzymes to break down complex waste into simple, high-value products | Bio-based product recovery from agro-food residues | El-Ramady et al. (2022) |
| Hydrothermal carbonization | Waste conversion into hydrochar under high temperature and pressure | Hydrochar production from rice husks | Jeevahan et al. (2021) |
| Advanced fermentation | Use of genetically engineered microbes to produce bioethanol or biobutanol | Bioethanol production from sugarcane waste in Brazil | Galanakis (2015) |

- Energy Utilization: Heat generated during combustion is used to produce steam, which drives turbines to generate electricity. The leftover ash is often landfilled or utilized in construction materials.

Recent trends have focused on addressing the challenges of traditional combustion methods, such as efficiency and environmental impact:

- Emission Control: Modern WTE plants use advanced technologies like scrubbers, filters, and catalytic converters to reduce harmful emissions like carbon dioxide ($CO_2$), nitrogen oxides ($NO_x$), and particulate matter (Pour & Makkawi, 2021).
- Co-combustion: Integrating food waste combustion with other processes like anaerobic digestion improves overall energy recovery. The digestate (solid residue) from digestion can be incinerated to extract additional energy (Zhang et al., 2022).

### 3.2.1 Integration with Circular Economy Principles

Combustion is increasingly aligned with the principles of a circular economy, aiming to maximize resource recovery and minimize waste. For example, the thermochemical pathways involve gasification and pyrolysis complementing traditional combustion, especially for food waste with high moisture content. These processes produce syngas and biochar, diversifying the energy outputs (Nayak & Bhushan, 2019). The lifecycle assessment is used in combustion facilities to evaluate their environmental footprint, ensuring that their operations are sustainable (Ingrao et al., 2018). Some benefits of combustion-based energy recovery include:

1. Volume Reduction: Food waste combustion significantly reduces the amount of waste sent to landfills, helping manage urban waste more effectively.
2. Energy Generation: Combustion produces renewable energy, which can power homes, industries, and even the WTE facility itself.
3. Environmental Benefits: Controlled combustion with modern emission-reduction technologies mitigates methane emissions that would otherwise occur in landfills.

These benefits are exemplified in real-world scenarios. The combustion of food waste for energy recovery has been implemented in various settings worldwide, showcasing its potential as a sustainable waste management and energy generation method. In Brazil, a thermochemical characterization study highlighted the energy recovery potential of food waste via combustion. The study emphasized its role in reducing waste volume and generating energy in urban settings (Gutierrez-Gomez et al., 2021). Aligning the combustion with circular economy principles, food waste can be integrated with other organic materials and burned in industrial plants to produce electricity and heat (Berenguer et al., 2023).

Hybrid systems in India combines combustion with regenerative dehydration, this approach improves efficiency by pre-drying wet food waste. This system is particularly effective in managing India's high-moisture food waste (Caton et al., 2010). China use integrated biorefineries, food waste digestate (residual solids from anaerobic digestion) is burned to extract additional energy, maximizing resource utilization (Zhang et al., 2022).

In Stockholm, food waste combustion is integrated with district heating systems, providing heat and electricity to local homes while reducing landfill dependency. In South Korea, advanced pyrolysis is combined with combustion to handle high-moisture food waste. This approach generates syngas, which is used for power generation in industrial facilities. In Denmark, a public–private partnership integrates food waste combustion with fertilizer production, ensuring complete utilization of ash by-products. In rural Kenya, small-scale combustion units convert food waste into energy for cooking and heating. These systems are vital in addressing energy poverty while managing organic waste locally.

### 3.2.2 Challenges, Limitations and Future Directions

Despite its benefits, combustion faces several challenges:

- Air Pollution: Even with advanced filters, incineration can release pollutants. Public concerns about air quality can lead to resistance against new WTE plants.
- Energy Balance: High moisture content in food waste reduces its calorific value, making combustion less efficient unless pre-drying technologies are employed.
- Ash Disposal: The ash generated requires proper management to avoid environmental contamination.

To address these challenges, researchers are exploring:

1. Hybrid Systems: Combining combustion with anaerobic digestion or gasification to maximize energy recovery.
2. Advanced Emission Controls: Developing technologies to capture carbon emissions and recycle them into useful by-products.
3. Decentralized Combustion Units: Smaller, localized units that minimize transportation costs and energy loss.

## 3.3 Biodiesel Production from Agro-Industrial Waste

Biodiesel production from agro-industrial waste is emerging as a vital approach to addressing environmental challenges while promoting energy sustainability. Biodiesel, a renewable and biodegradable fuel, is produced from organic materials such as vegetable oils, animal fats, or waste cooking oils. It offers a cleaner alternative to fossil fuels by significantly reducing greenhouse gas emissions.

Agro-industrial waste, generated from agricultural and industrial processes, provides an abundant and cost-effective feedstock for biodiesel production. These wastes include residues like rice husks, sugarcane bagasse, fruit pomace, and dairy scum oil, which are rich in lipids, cellulose, and hemicellulose, making them ideal for biofuel production (Maleki et al., 2024; Mekunye & Makinde, 2024).

The production process involves converting these wastes into biodiesel through transesterification, where triglycerides react with alcohol in the presence of catalysts to yield biodiesel and glycerol. Advanced methods using heterogeneous catalysts derived from agricultural waste, such as rice husk ash, are gaining popularity for their cost efficiency and environmental benefits (Maleki et al., 2024). Moreover, biocatalysis employing enzymes like lipase offers an eco-friendly alternative, particularly for feedstocks with high free fatty acids (Ribeiro et al., 2024). These technological advancements have significantly enhanced the feasibility and efficiency of biodiesel production from waste materials (Fig. 3.2).

The advantages of this approach extend beyond energy production. Economically, it reduces the costs associated with biodiesel production and waste management while creating additional revenue streams for industries (Mekunye & Makinde, 2024). Environmentally, repurposing waste for biodiesel mitigates pollution and promotes a circular economy, turning waste into valuable resources (Oñate et al., 2024). Furthermore, the

**Fig. 3.2** Methods in biodiesel production

integration of biodiesel production into biorefineries supports the simultaneous generation of co-products like bioethanol, biogas, and biochar, maximizing resource utilization (Sampaio et al., 2024).

Despite its potential, the field faces challenges such as feedstock variability, contamination, and high initial investment costs. Addressing these issues has led to innovations such as green catalysts derived from waste materials, integrated biorefineries, and the application of machine learning for process optimization (Maleki et al., 2024; Ribeiro et al., 2024). Microbial and enzymatic pathways are also being explored to preprocess waste, improving biodiesel yield and reducing costs (Miranda-Sosa et al., 2024).

Case studies demonstrate the practicality of this approach globally. In Southeast Asia, palm oil mill effluent is a key feedstock (Mekunye & Makinde, 2024), while rice husk and sugarcane bagasse are extensively utilized in India and Brazil, respectively (Maleki et al., 2024; Oñate et al., 2024). These examples showcase how agro-industrial residues can support biodiesel production while addressing local environmental challenges. In the United States, dairy waste scum is transformed into biodiesel, showcasing innovative solutions for industrial waste management (Maleki et al., 2024).

The future of biodiesel production from agro-industrial waste is promising, with support from government policies and advancements in technology. Incentives like subsidies and renewable energy mandates encourage adoption, while innovations in nanotechnology and artificial intelligence optimize production processes (Ribeiro et al., 2024). Sustainability metrics, such as life cycle assessments, are increasingly used to evaluate environmental impacts, ensuring the development of eco-friendly biodiesel systems (Sampaio et al., 2024). Developing nations are leading efforts to utilize agro-industrial waste, addressing energy needs and waste management simultaneously (Miranda-Sosa et al., 2024).

Biodiesel production from agro-industrial waste stands at the intersection of environmental sustainability, energy security, and technological innovation. With continued research and policy support, this approach has the potential to revolutionize the energy landscape and contribute significantly to a cleaner, greener future.

## 3.4 Corn Waste to Gasoline

The transformation of corn waste, specifically corn stover, into gasoline is a revolutionary advancement in renewable energy research. As global energy demands continue to rise, the importance of transitioning from fossil fuels to sustainable alternatives has never been greater. Corn stover, the leaves, stalks, and cobs left after corn harvest, emerges as a promising feedstock due to its abundance and high lignocellulosic content. This process not only offers a solution to agricultural waste management but also aligns with efforts to reduce greenhouse gas emissions and mitigate the environmental impacts of fossil fuel use (Avendano et al., 2024; Shi et al., 2024).

Corn stover is one of the most widely available agricultural residues in corn-producing regions such as the United States, Brazil, and China. Each year, millions of tons of corn stover remain in fields, often left to decompose or burned, contributing to environmental pollution. Rich in cellulose, hemicellulose, and lignin, corn stover serves as an excellent raw material for biofuel production (Shi et al., 2024). Its utilization in the production of biofuels, including gasoline, addresses dual challenges: managing agricultural waste and reducing dependence on non-renewable resources.

Efforts to harness corn stover as a feedstock involve strategies to optimize its collection, storage, and pre-treatment. Pre-treatment methods such as steam explosion, acid hydrolysis, and enzymatic hydrolysis play a crucial role in breaking down the complex lignocellulose structure into fermentable sugars. These processes improve the efficiency of downstream biochemical and thermochemical conversion technologies, ensuring higher yields of biofuels (Xu et al., 2024).

### 3.4.1 Technological Pathways for Conversion

The conversion of corn stover into gasoline involves two primary technological pathways: thermochemical and biochemical processes. Each pathway offers unique advantages and challenges, often complemented by innovations in catalytic systems.

Thermochemical Processes: Thermochemical conversion methods include gasification and pyrolysis. Gasification involves heating corn stover in a controlled oxygen environment to produce syngas, a mixture of carbon monoxide and hydrogen. This syngas can then be processed through Fischer–Tropsch synthesis to produce synthetic gasoline. Advanced systems integrate water–gas shift reactions to enhance the hydrogen content, improving overall fuel yield (Kumar et al., 2024).

Pyrolysis, another widely used method, thermally decomposes corn stover into bio-oil, biochar, and syngas in the absence of oxygen. Bio-oil, a key product, can be upgraded into gasoline-range hydrocarbons using catalytic hydrodeoxygenation. Recent advancements in co-pyrolysis, where corn stover is processed with materials like plastics or rubber, have shown promise in improving the quality and yield of liquid fuels. Catalysts such as zeolites and Ni-based systems are integral to these processes, enhancing selectivity and efficiency (Kohler et al., 2024).

Biochemical Processes: The biochemical conversion of corn stover involves the enzymatic breakdown of cellulose into glucose, which is then fermented into ethanol (Fig. 3.3). Ethanol serves as a precursor for gasoline production through catalytic processes. Biochemical methods benefit from ongoing research into enzyme efficiency and cost reduction. Furthermore, the integration of these methods into biorefineries allows for the co-production of other bio-based products, enhancing economic viability (Dussan et al., 2025).

## 3.4 Corn Waste to Gasoline

**Fig. 3.3** General diagram for the conversion of corn feedstock to ethanol

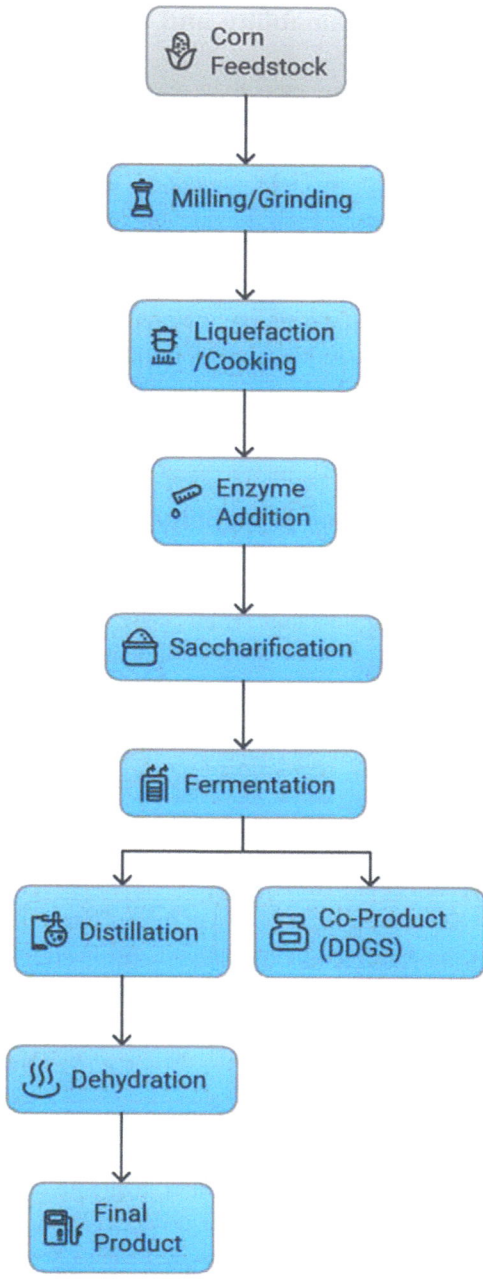

## 3.4.2 Sustainability and Environmental Benefits

One of the most compelling reasons for pursuing corn stover-to-gasoline technologies is the significant environmental benefits they offer. By utilizing agricultural residues, these technologies reduce the need for fossil fuel extraction, thereby minimizing greenhouse gas emissions. Studies consistently show that biofuels derived from corn stover emit significantly less carbon dioxide compared to conventional gasoline (Shi et al., 2024; Sun et al., 2024).

The integration of biorefineries amplifies these benefits. In addition to gasoline, biorefineries produce valuable by-products such as biochar, which can be used as a soil amendment to sequester carbon and improve soil health. Heat generated during conversion processes can be reused within the facility, further reducing the overall carbon footprint (Abedin et al., 2024).

## 3.4.3 Economic Viability and Challenges

Despite their potential, the economic viability of corn stover-to-gasoline systems remains a critical consideration. High capital costs associated with biorefinery infrastructure, coupled with feedstock variability, present significant challenges. Variations in corn stover composition can affect conversion efficiency, necessitating the development of adaptable processing technologies (Xu et al., 2024).

To address these challenges, researchers are focusing on cost reduction strategies, such as developing low-cost catalysts and improving process integration. Catalysts derived from waste materials and advanced systems like microwave-assisted pyrolysis are being explored for their potential to enhance energy efficiency and lower production costs (Abedin et al., 2024).

The future of corn stover-to-gasoline production lies in the integration of advanced technologies and policy support. Artificial intelligence and machine learning are increasingly being employed to optimize conversion processes, predict performance, and reduce inefficiencies. Hybrid pathways that combine gasification and biochemical conversion offer flexibility in product outputs, catering to diverse market demands (Wang et al., 2024).

Policy frameworks play a critical role in driving the adoption of these technologies. Government incentives such as renewable energy mandates, carbon credits, and subsidies for biofuel production create a favorable environment for investment and innovation. In regions with abundant corn production, such as the U.S., China, and Brazil, large-scale projects are being implemented to showcase the feasibility of these systems.

The global adoption of corn stover-to-gasoline technologies holds significant implications for energy security and environmental sustainability. By repurposing agricultural waste into valuable energy resources, these technologies contribute to a circular economy, where waste is minimized, and resources are used efficiently. Moreover, the scalability of these systems makes them an attractive option for addressing energy needs in both developed and developing countries.

Corn waste-to-gasoline technologies represent a critical innovation in the transition to renewable energy. By leveraging advanced conversion methods, catalytic innovations, and integrated biorefineries, this approach addresses pressing challenges in waste management, energy production, and environmental sustainability. While economic and technical challenges remain, ongoing research and supportive policies are paving the way for widespread adoption. As the world moves toward a greener future, corn stover-to-gasoline systems are poised to play a key role in meeting global energy demands while reducing the carbon footprint of transportation fuels.

## 3.5 Biodiesel Production from Waste Cooking Oil

The global demand for sustainable energy sources has prompted a substantial interest in biodiesel production from waste cooking oil (WCO). WCO, a byproduct of the food industry, is an attractive feedstock due to its abundance and cost-effectiveness compared to virgin oils. Its utilization not only addresses the challenges associated with the disposal of used cooking oil but also supports the principles of circular economies by converting waste into a valuable resource. This approach contributes to the reduction of environmental pollution and the mitigation of fossil fuel dependency. Research has repeatedly demonstrated that biodiesel derived from WCO meets international fuel quality standards and provides substantial environmental benefits. For instance, research has demonstrated that biodiesel leads to significant reductions in greenhouse gas emissions when compared to traditional fossil fuels (Goh et al., 2020; Singh et al., 2019).

The collection of WCO has emerged as a critical step in biodiesel production, requiring collaboration among local authorities, restaurants, and households. Notably, restaurants and fast-food chains are considered particularly valuable sources of WCO due to the substantial volumes they generate. To ensure a reliable and consistent supply of WCO, it is imperative to optimize collection networks and incentivize participation. The implementation of localized collection systems has been demonstrated to reduce logistical challenges and promote community involvement, thereby fostering the adoption of sustainable energy practices.

### 3.5.1 Advancements in Biodiesel Production Processes

Technological advancements in biodiesel production have considerably augmented the viability of utilizing WCO as a feedstock (Fig. 3.4). A notable breakthrough has been the

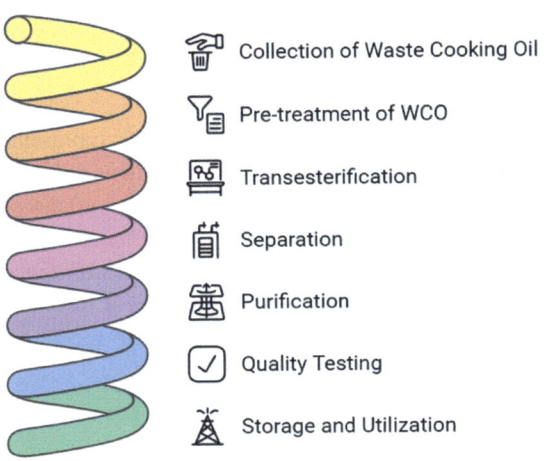

**Fig. 3.4** General overview of the production process of biodiesel from WCO

optimization of the transesterification process, which involves the conversion of triglycerides present in oil into biodiesel and glycerol through a chemical reaction. A plethora of catalysts, encompassing acid, base, and enzymatic options, have been thoroughly investigated to enhance the efficiency and yield of the reaction. Among the various catalysts employed, base catalysts, such as sodium hydroxide, have gained prominence due to their cost-effectiveness and high efficiency. Nonetheless, acid catalysts have emerged as a more effective solution for oils with high free fatty acid content, a characteristic frequently observed in WCO. Recent studies have identified the potential of enzymatic catalysts, which offer enhanced specificity and reduced energy requirements. However, their cost remains a significant challenge (Singh et al., 2021).

Another critical area of innovation is the pre-treatment of WCO to remove impurities such as water and food residues. The implementation of effective pre-treatment methodologies is paramount in ensuring the consistency of fuel quality and extending the lifespan of catalysts. The development of advanced filtration and degumming techniques has led to a significant enhancement in the efficiency of the pre-treatment process. Research has demonstrated that these methodologies substantially improve the overall quality of biodiesel, thereby ensuring its compliance with international standards for viscosity, cetane number, and other properties deemed essential for engine compatibility (Gouran et al., 2021).

### 3.5.2 Policy and Regulatory Support

The role of policy and regulation in promoting the use of WCO for biodiesel production cannot be overstated. Numerous governments worldwide have instituted measures to promote the collection and processing of WCO. These measures encompass a range of policy instruments, including tax incentives for biodiesel producers, subsidies for WCO collection programs, and penalties for illegal disposal of used cooking oil. The implementation of such policies has been demonstrated to have a multifaceted impact, encompassing both the mitigation of environmental degradation and the cultivation of economic prospects. This is achieved by establishing a market for WCO-based biodiesel, thereby creating opportunities for economic growth and development.

In certain regions, collaborations between governmental bodies and private enterprises have played a pivotal role in the establishment of efficient WCO collection systems. For instance, collaborations with restaurants and fast-food chains have ensured a steady supply of feedstock for biodiesel production. Furthermore, public awareness campaigns emphasizing the environmental and economic benefits of WCO recycling have encouraged increased participation from households. Furthermore, SWOT analyses of biodiesel production systems have identified robust policy frameworks as a key factor for success (Liu et al., 2018).

### 3.5.3 Economic Viability and Applications

The economic viability of biodiesel production from waste cooking oil (WCO) is a significant factor in its adoption, particularly in regions with limited access to conventional fuels. Research has demonstrated that localized biodiesel production is not only cost-effective but also beneficial for energy security. The utilization of locally available waste resources has been shown to reduce a community's reliance on imported fuels, thereby creating employment opportunities in the collection, processing, and distribution of biodiesel (Park et al., 2019).

Beyond its use as a transportation fuel, WCO-derived biodiesel has demonstrated potential in other applications. For instance, it can be utilized as a feedstock for biojet fuel production, thereby offering a renewable alternative for the aviation industry. Furthermore, the potential of WCO in the generation of electricity and other forms of renewable energy has been investigated. The aforementioned applications underscore the versatility of WCO as a resource and highlight its value in promoting energy sustainability (Goh et al., 2020; Panadare & Rathod, 2015).

### 3.5.4 Challenges and Research Directions

Notwithstanding its considerable potential, the production of biodiesel from WCO is encumbered by several challenges. A primary concern pertains to the variability in the quality of WCO, which has the capacity to influence the efficiency of the transesterification process and the characteristics of the final product. In addressing this challenge, researchers have directed their efforts towards the development of advanced analytical methods for evaluating WCO quality and the design of flexible production systems capable of adapting to variations in feedstock. However, the high cost of enzymatic catalysts and pre-treatment processes continues to impede large-scale implementation.

Another significant challenge is the establishment of efficient and reliable WCO collection networks. While collaborative endeavors with restaurants and food industries have yielded positive outcomes in certain regions, broader engagement from households is imperative to address the escalating demands for feedstock. The implementation of public awareness campaigns and the provision of incentives for household participation are identified as pivotal in this regard.

Subsequent research endeavors should investigate the incorporation of co-products from biodiesel production, such as glycerol, into other industrial applications. This approach would enhance the economic viability of the production process and align with the principles of waste valorization. Furthermore, state-of-the-art bibliometric analyses have identified emergent research trends in catalyst development and lifecycle assessments, which are imperative for enhancing the sustainability and scalability of biodiesel production systems (Chen et al., 2021).

The adoption of WCO as a biodiesel feedstock exemplifies the intersection of waste management and renewable energy, fostering sustainability. By reducing environmental pollution, lowering greenhouse gas emissions, and supporting local economies, biodiesel derived from WCO contributes to the achievement of global sustainability goals. The incorporation of WCO-based biodiesel into national energy strategies has the potential to substantially enhance energy resilience and reduce reliance on non-renewable resources.

In order to facilitate the widespread adoption of WCO-derived biodiesel, it is imperative to address the technical, economic, and social barriers to its implementation. This necessitates the allocation of resources to research and development initiatives aimed at enhancing production technologies, the establishment of international collaborations to facilitate the exchange of best practices, and the implementation of comprehensive policy measures to provide the necessary support to the industry. By leveraging these strategies, biodiesel production from WCO can play a pivotal role in the transition to a more sustainable and inclusive energy future (Singh et al., 2019; Yaakob et al., 2013).

In summary, biodiesel production from WCO is a sustainable, economically viable and environmentally friendly alternative to conventional fossil fuels. The viability of this approach hinges on the sustained advancement of production technologies, the establishment of supportive policy frameworks, and the active engagement of all stakeholders in the collection and recycling of WCO. Through collaborative endeavors and the development of innovative solutions, this promising field has the potential to contribute to the transformation of the energy landscape towards greater sustainability and cleanliness.

# References

Abedin, A., Bai, X., & Muley, P. (2024). *From waste to clean fuel: Using microwave chemistry to achieve process decarbonization*. American Chemical Society Fall 2024 National Meeting.

Avendano, M. E., Lao, J., Fu, Q., Nair, S., & Realff, M. J. (2024). Environmental impact of Simulated Moving Bed (SMB) on the recovery of 2,3-butanediol on an integrated biorefinery. *LAPSE Living Archive for Process Systems Engineering, 3*, 660–667. https://doi.org/10.69997/sct.121375

Berenguer, C. V., Perestrelo, R., Pereira, J. A. M., & Câmara, J. S. (2023). Management of agri-food waste based on thermochemical processes towards a circular bioeconomy concept: The Case study of the Portuguese industry. *Processes, 11*(10), 2–24. https://doi.org/10.3390/pr11102870

Borja, R., & Fernández-Rodríguez, M. J. (2021). Chapter 7—Energy recovery as added value from food and agricultural solid wastes. In S. Kumar, R. Kumar, & A. Pandey (Eds.), *Current developments in biotechnology and bioengineering strategic perspectives in solid waste and wastewater management* (pp. 151–174). Elsevier. https://doi.org/10.1016/B978-0-12-821009-3.00007-5

Caton, P. A., Carr, M. A., Kim, S. S., & Beautyman, M. J. (2010). Energy recovery from waste food by combustion or gasification with the potential for regenerative dehydration: A case study. *Energy Conversion and Management, 51*(6), 1157–1169. https://doi.org/10.1016/j.enconman.2009.12.025

Cecconet, D., Molognoni, D., Callegari, A., & Capodaglio, A. G. (2018). Agro-food industry wastewater treatment with microbial fuel cells: Energetic recovery issues. *International Journal of Hydrogen Energy, 43*(1), 500–511. https://doi.org/10.1016/j.ijhydene.2017.07.231

Chen, C., Chitose, A., Kusadokoro, M., Nie, H., Xu, W., Yang, F., & Yang, S. (2021). Sustainability and challenges in biodiesel production from waste cooking oil: An advanced bibliometric analysis. *Energy Reports, 7*, 4022–4034. https://doi.org/10.1016/j.egyr.2021.06.084

Dussan, K., Hoek, M., de Vrije, T., van de Vondervoort, R., Bonouvrie, P., Caliskan, R., Parenti, A., Zegada-Lizarazu, W., Monti, A., Smit, A. T., & López-Contreras, A. M. (2025). Performance of mild acetone organosolv fractionation on lignocellulosic feedstocks from new cropping systems for production of advanced bioethanol. *Industrial Crops and Products, 223*, 120156. https://doi.org/10.1016/j.indcrop.2024.120156

Economou, F., Voukkali, I., Papamichael, I., Phinikettou, V., Loizia, P., Naddeo, V., Sospiro, P., Liscio, M. C., Zoumides, C., Țîrcă, D. M., & Zorpas, A. A. (2024). Turning food loss and food waste into watts: A review of food waste as an energy source. *Energies, 17*(13), 1–30. https://doi.org/10.3390/en17133191

El-Ramady, H., Brevik, E. C., Bayoumi, Y., Shalaby, T. A., El-Mahrouk, M. E., Taha, N., Elbasiouny, H., Elbehiry, F., Amer, M., Abdalla, N., Prokisch, J., Solberg, S., & Ling, W. (2022). An overview of agro-waste management in light of the water-energy-waste nexus. *Sustainability (Switzerland), 14*(23), 1–30. https://doi.org/10.3390/su142315717

Galanakis, C. M. (2015). *Food waste recovery: Processing technologies and industrial techniques*. Academic Press.

Goh, B. H. H., Chong, C. T., Ge, Y., Ong, H. C., Ng, J.-H., Tian, B., Ashokkumar, V., Lim, S., Seljak, T., & Józsa, V. (2020). Progress in utilisation of waste cooking oil for sustainable biodiesel and biojet fuel production. *Energy Conversion and Management, 223*, 113296. https://doi.org/10.1016/j.enconman.2020.113296

Gouran, A., Aghel, B., & Nasirmanesh, F. (2021). Biodiesel production from waste cooking oil using wheat bran ash as a sustainable biomass. *Fuel, 295*, 120542. https://doi.org/10.1016/j.fuel.2021.120542

Gutierrez-Gomez, A. C., Gallego, A. G., Palacios-Bereche, R., Tofano de Campos Leite, J., & Pereira Neto, A. M. (2021). Energy recovery potential from Brazilian municipal solid waste via combustion process based on its thermochemical characterization. *Journal of Cleaner Production, 293*, 126145. https://doi.org/10.1016/j.jclepro.2021.126145

Ingrao, C., Faccilongo, N., Di Gioia, L., & Messineo, A. (2018). Food waste recovery into energy in a circular economy perspective: A comprehensive review of aspects related to plant operation and environmental assessment. *Journal of Cleaner Production, 184*, 869–892. https://doi.org/10.1016/j.jclepro.2018.02.267

Jeevahan, J., Anderson, A., Sriram, V., Durairaj, R. B., Britto Joseph, G., & Mageshwaran, G. (2021). Waste into energy conversion technologies and conversion of food wastes into the potential products: A review. *International Journal of Ambient Energy, 42*(9), 1083–1101. https://doi.org/10.1080/01430750.2018.1537939

Kohler, A. J., Vincent Sahayaraj, D., Ehlers, A., Bai, X., Shanks, B. H., & Tessonnier, J.-P. (2024). Nonreductive fast lignin solvolysis in flow-through reactors for the sustainable production of BTEX aromatics from corn stover. *ACS Sustainable Chemistry & Engineering, 12*(33), 12504–12515. https://doi.org/10.1021/acssuschemeng.4c03722

Kumar, A., Bell, D. C., Yang, Z., Heyne, J., Santosa, D. M., Wang, H., Zuo, P., Wang, C., Mittal, A., Klein, D. P., Manto, M. J., Chen, X., & Yang, B. (2024). A simultaneous depolymerization and hydrodeoxygenation process to produce lignin-based jet fuel in continuous flow reactor. *Fuel Processing Technology, 263*, 108129. https://doi.org/10.1016/j.fuproc.2024.108129

Liu, Y., Liu, T., Agyeiwaa, A., & Li, Y. (2018). A SWOT analysis of biodiesel production from waste cooking oil. *IOP Conference Series: Earth and Environmental Science, 170*(2), 22136. https://doi.org/10.1088/1755-1315/170/2/022136

Maleki, B., Esmaeili, H., Venkatesh, Y. K., & Amruth, E. (2024). Valorization of dairy waste scum oil and rice husk ash-supported CuO nanocatalyst towards cleaner production of biodiesel: A waste-to-energy approach. *Process Safety and Environmental Protection, 192*, 1393–1407. https://doi.org/10.1016/j.psep.2024.10.124

Mekunye, F., & Makinde, P. (2024). Sustainable biofuel production from agricultural waste: Advances in biochemical and thermochemical conversion pathways. In M. Abdel-Raheem (Ed.), *Current research progress in agricultural sciences Vol. 6* (Issue SE-Chapters, pp. 117–139). https://doi.org/10.9734/bpi/crpas/v6/3285

Miranda-Sosa, A., del Moral, S., Infanzón-Rodriguez, M. I., & Aguilar-Uscanga, M. G. (2024). Reappraisal of different agro-industrial waste for the optimization of cellulase production from Aspergillus niger ITV02 in a liquid medium using a Box–Benkhen design. *3 Biotech, 14*(11), 278. https://doi.org/10.1007/s13205-024-04120-5

Nayak, A., & Bhushan, B. (2019). An overview of the recent trends on the waste valorization techniques for food wastes. *Journal of Environmental Management, 233*, 352–370. https://doi.org/10.1016/j.jenvman.2018.12.041

Oñate, A., Travieso Pedroso, D., Valenzuela, M., Blanco Machin, E., & Tuninetti, V. (2024). Production of high-calorific hybrid biofuel pellets from urban plastic waste and agro-industrial by-products. *Journal of Cleaner Production, 479*, 144046. https://doi.org/10.1016/j.jclepro.2024.144046

Panadare, D. C., & Rathod, V. K. (2015). Applications of waste cooking oil other than biodiesel: A review. *Iranian Journal of Chemical Engineering (IJChE), 12*(3), 55–76. https://www.ijche.com/article_11253.html

Park, S. H., Khan, N., Lee, S., Zimmermann, K., DeRosa, M., Hamilton, L., Hudson, W., Hyder, S., Serratos, M., Sheffield, E., Veludhandi, A., & Pursell, D. P. (2019). Biodiesel production from locally sourced restaurant waste cooking oil and grease: Synthesis, characterization, and performance evaluation. *ACS Omega, 4*(4), 7775–7784. https://doi.org/10.1021/acsomega.9b00268

Pour, F. H., & Makkawi, Y. T. (2021). A review of post-consumption food waste management and its potentials for biofuel production. *Energy Reports, 7*, 7759–7784. https://doi.org/10.1016/j.egyr.2021.10.119

Ramanaiah, S. V, Chandrasekhar, K., Cordas, C. M., & Potoroko, I. (2023). Bioelectrochemical systems (BESs) for agro-food waste and wastewater treatment, and sustainable bioenergy—A review. *Environmental Pollution, 325*, 121432. https://doi.org/10.1016/j.envpol.2023.121432

Ribeiro, T. S., de Araújo Sobrinho, I., Gonçalves, M. A., da Silva Lima, V., Figueira, B. A. M., da Rocha Filho, G. N., & da Conceição, L. R. V. (2024). Green synthesis of biodiesel from magnetic basic biochar derived from Amazonian murici residual biomass: Optimization, kinetic, thermodynamic, and environmental studies. *Journal of Environmental Chemical Engineering, 12*(6), 114725. https://doi.org/10.1016/j.jece.2024.114725

Sampaio, I. C. F., Silva, F. N., de Moura, I. V. L., Gonçalves, M. S., Franco, M., & de Almeida, P. F. (2024). Sustainable horizons: Navigating challenges in butanol production from lignocellulosic by-products BT. In C. A. Taft & S. R. de Lazaro (Eds.), *Progress in hydrogen energy, fuel cells, nano-biotechnology and advanced, bioactive compounds. Engineering materials* (pp. 401–415). Springer Nature Switzerland. https://doi.org/10.1007/978-3-031-75984-0_17

Shi, J., Luo, Z., Sun, H., Qian, Q., Wei, Q., & Li, L. (2024). Enhancing corn stover to bio-jet fuel process: Valorizing lignin-enriched residue for energy, economic, and environmental benefits. *Biomass and Bioenergy, 188*, 107338. https://doi.org/10.1016/j.biombioe.2024.107338

Singh, D., Sharma, D., Soni, S. L., Sharma, S., & Kumari, D. (2019). Chemical compositions, properties, and standards for different generation biodiesels: A review. *Fuel, 253*, 60–71. https://doi.org/10.1016/j.fuel.2019.04.174

Singh, D., Sharma, D., Soni, S. L., Inda, C. S., Sharma, S., Sharma, P. K., & Jhalani, A. (2021). A comprehensive review of biodiesel production from waste cooking oil and its use as fuel in compression ignition engines: 3rd generation cleaner feedstock. *Journal of Cleaner Production, 307*, 127299. https://doi.org/10.1016/j.jclepro.2021.127299

Sun, J., Zhang, Z., Liu, J., & Zhang, S. (2024). Experimental study on biogas fermentation of corn stover pretreated with compound microbial agent. *Energy, 306*, 132469. https://doi.org/10.1016/j.energy.2024.132469

Wang, F., Jiang, X., Liu, Y., Zhang, G., Zhang, Y., Jin, Y., Shi, S., Men, X., Liu, L., Wang, L., Liao, W., Chen, X., Chen, G., Liu, H., Ahmad, M., Fu, C., Wang, Q., Zhang, H., & Lee, S. Y. (2024). Tobacco as a promising crop for low-carbon biorefinery. In *The innovation* (vol. 5, Issue 5). Elsevier. https://doi.org/10.1016/j.xinn.2024.100687

Xu, W., Zhang, J., Wu, Q., Wang, Y., Zhao, W., Zhu, Z., Wang, Y., & Cui, P. (2024). Energy, exergy and economic (3E) analyses of a novel DME-power polygeneration system with $CO_2$ capture based on biomass gasification. *Applied Energy, 374*, 124031. https://doi.org/10.1016/j.apenergy.2024.124031

Yaakob, Z., Mohammad, M., Alherbawi, M., Alam, Z., & Sopian, K. (2013). Overview of the production of biodiesel from Waste cooking oil. *Renewable and Sustainable Energy Reviews, 18*, 184–193. https://doi.org/10.1016/j.rser.2012.10.016

Zhang, L., Yang, P., Zhu, K., Ji, X., Ma, J., Mu, L., Ullah, F., Ouyang, W., & Li, A. (2022). Biorefinery-oriented full utilization of food waste and sewage sludge by integrating anaerobic digestion and combustion: Synergistic enhancement and energy evaluation. *Journal of Cleaner Production, 380*, 134925. https://doi.org/10.1016/j.jclepro.2022.134925

# Biogas Production from Food Waste

## 4.1 Introduction

Biogas production is a process that converts organic materials into a mixture of gases, primarily methane ($CH_4$) and carbon dioxide ($CO_2$), through the action of microorganisms in the absence of oxygen. This process, known as anaerobic digestion (AD), has garnered increasing recognition as a sustainable method for generating renewable energy and addressing global waste management challenges. Food waste, given its high organic content, offers considerable potential for biogas production. The transformation of this frequently discarded resource into energy and nutrient-rich by-products is a critical aspect of biogas systems. These systems contribute to reducing greenhouse gas emissions, mitigating landfill overflows, and creating a circular economy. The biogas produced can be utilized for heating, electricity generation, or upgraded to biomethane for transportation, making it a versatile and valuable energy source (Weiland, 2010).

Biogas production from food waste involves a stepwise anaerobic digestion process, carried out by specific microorganisms. The process commences with hydrolysis, wherein complex molecules such as carbohydrates, proteins, and fats are broken down into simpler compounds, including sugars, amino acids, and fatty acids. This stage establishes the foundation for further microbial activity; however, it is often the slowest, requiring effective pretreatment to enhance its efficiency. Subsequently, during acidogenesis, the simpler compounds are converted into volatile fatty acids, alcohols, hydrogen, and carbon dioxide. These intermediates are subsequently subjected to further breakdown in acetogenesis, resulting in the formation of acetic acid, hydrogen, and carbon dioxide. These elements then undergo a transformation into methane and carbon dioxide by methanogenic microorganisms during methanogenesis. The efficiency and methane yield of the system are determined by this final stage, which is critical for energy output (Pramanik et al., 2019).

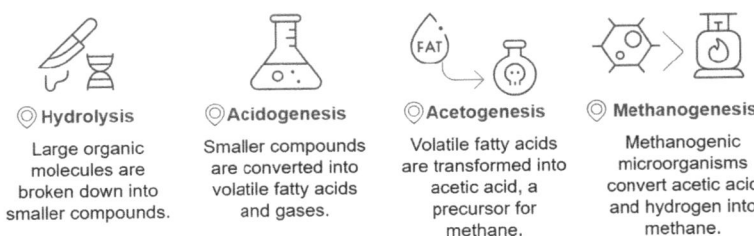

**Fig. 4.1** Key stages of anaerobic digestion to produce biogas

Optimal conditions are imperative for maximizing biogas production. Specifically, mesophilic temperatures (30–40 °C) or thermophilic conditions (50–60 °C) have been shown to provide an optimal environment for microbial activity. Furthermore, the maintenance of a neutral to slightly alkaline pH range (6.8–7.5) is imperative for the optimal proliferation of methanogens. Furthermore, it is imperative to maintain a balanced organic loading rate and ensure sufficient retention time of 15–30 days to avoid overloading the system and ensure complete digestion (Ferdeș et al., 2022) (Fig. 4.1).

Food waste is particularly suitable for biogas production because it contains high levels of easily digestible organic materials, such as carbohydrates, fats, and proteins. In comparison with other substrates, such as agricultural residues or manure, food waste has been shown to yield significantly higher methane, thereby making it an efficient feedstock for anaerobic digestion (Mirmohamadsadeghi et al., 2019). Despite its promise, the production of biogas from food waste is encumbered by challenges that must be addressed for wider adoption. Feedstock variability poses a significant challenge, as food waste composition is subject to frequent changes, resulting in inconsistent biogas yields. Furthermore, food waste frequently contains inhibitors such as salt, grease, or preservatives, which can impede microbial activity and reduce methane production. The implementation of effective pretreatment strategies is imperative to mitigate these issues (Pramanik et al., 2019).

The biogas industry is undergoing rapid advancements, particularly in the treatment and processing of food waste. A notable area of research interest is the development of pretreatment technologies, which aim to enhance the digestibility of food waste. The recycling of food waste into biogas offers a multitude of benefits. It curtails the volume of waste destined for landfills, thereby mitigating uncontrolled methane emissions that contribute to climate change. Moreover, the process of anaerobic digestion generates digestate, a nutrient-rich by-product that has the potential to substitute for synthetic fertilizers, thereby completing the cycle of nutrient regeneration and reducing the reliance on chemical inputs (Ferdeș et al., 2022). Co-digestion represents a growing trend in the field. In this process, food waste is integrated with other organic materials, such as manure or crop residues, within biogas systems. This integration serves to optimize the nutrient profiles, mitigate inhibitor effects, and enhance overall gas production. For instance, the co-digestion process with manure introduces essential microorganisms that accelerate the digestion process, while concurrently balancing pH levels and reducing the risk of system instability (Alghoul et al., 2019).

The economic feasibility of such systems is being enhanced by the development of smaller, community-level biogas plants that process food waste at the local level. These systems offer distinct advantages, including the reduction of transportation expenses and the provision of renewable energy to neighboring households or businesses. Lifecycle assessments have demonstrated that biogas production systems have a reduced environmental impact in comparison with conventional waste disposal methods, such as landfilling. This attribute renders them a preferred solution for waste-to-energy conversion (Jin et al., 2015). Economic and regulatory barriers also pose significant hurdles. The initial costs associated with the establishment of a biogas plant can be substantial, particularly for small-scale operators. Additionally, the inconsistency in government policies concerning renewable energy subsidies can have a deleterious effect on the financial viability of these projects (Al-Wahaibi et al., 2020).

The future of biogas production is contingent upon innovation and integration. Advancements in gas upgrading technologies have facilitated the purification of biogas into biomethane, which is then able to be injected into natural gas grids or utilized as vehicle fuel. In urban areas, the implementation of decentralized biogas plants has the potential to process food waste at the local level, thereby reducing transportation-related emissions and providing communities with a sustainable energy source. Governments are progressively acknowledging the promise of biogas systems and are thus implementing incentives to encourage their adoption, including feed-in tariffs and carbon credits (Mirmohamadsadeghi et al., 2019).

The production of biogas from food waste signifies a transformative approach to waste management and renewable energy generation. Biogas systems have been developed for the purpose of converting discarded food into energy and nutrient-rich fertilizer. This represents a transformative approach to waste management and renewable energy generation, as it addresses pressing environmental issues.

## 4.2 Biotransformation of Food Waste into Biogas

Biotransformation of food waste into biogas is a transformative process that addresses two critical global challenges: waste management and renewable energy generation. Biogas production through anaerobic digestion (AD) is an efficient way to convert food waste into energy, leveraging microbial activity to break down organic matter in the absence of oxygen. This process produces biogas, primarily composed of methane ($CH_4$) and carbon dioxide ($CO_2$), which can be used for electricity, heating, and even as a vehicle fuel. A nutrient-rich by-product called digestate is also generated, which serves as an organic fertilizer. Recent advancements in biotransformation techniques, integration with the circular economy, and emerging technologies have significantly improved the efficiency and scalability of this process.

## 4.2.1 Advances in Biotransformation Techniques and Optimization

Pretreatment of food waste constitutes a critical step in the biotransformation process, as it prepares the substrate for efficient microbial digestion. Food waste frequently contains complex organic molecules, including cellulose, lignin, proteins, and fats, which exhibit resistance to degradation. Pretreatment methods have been developed to address these challenges by disrupting the molecular structures of these recalcitrant materials, thereby enhancing their accessibility to microbial enzymes and accelerating the anaerobic digestion process. A range of pretreatment techniques were mentioned by Ferdeș et al., (2022), each offering distinct advantages.

- Physical Pretreatment: Physical methods encompass grinding, shredding, and thermal treatment. These techniques have been demonstrated to increase the surface area of food waste, thereby facilitating enhanced microbial access to organic material. For instance, thermal pretreatment, which involves heating food waste to disrupt cell walls, has been shown to enhance the effectiveness of enzymes and microbes in breaking down the substrate. This approach has been demonstrated to enhance biogas yields to a considerable extent. Research findings have demonstrated that the mechanical treatment of food waste using a Hollander beater has been shown to increase methane production by up to 80% compared to untreated waste.
- Chemical Pretreatment: Chemical methods employ acids, alkalis, or oxidizing agents to degrade recalcitrant organic compounds, such as lignin. For instance, sodium hydroxide is frequently employed to enhance the solubility of fibrous, plant-based food waste, thereby facilitating accelerated microbial degradation. However, judicious application is imperative to circumvent the introduction of toxic by-products into the system.
- Biological Pretreatment: This eco-friendly approach utilizes enzymes or specific microbes to initiate the breakdown of organic material prior to anaerobic digestion. Enzymes such as cellulase and protease target specific components of food waste, such as carbohydrates and proteins, respectively. The adoption of biological pretreatment is increasing due to its sustainability and its capacity to function effectively in large-scale systems.

In addition to pretreatment, the optimization of operational parameters of anaerobic digestion is imperative for the maximization of biogas production. The efficiency of the process is influenced by several key factors.

The pH level is a critical factor in the methane production process because methanogens, the microbes responsible for methane production, thrive in a pH range of 6.8–7.5. It is imperative to maintain this range to prevent acidification of the digester, as this can lead to inhibition of microbial activity and subsequent reduction in methane yields. It is imperative to ensure sufficient retention time, which is typically between 15 and 30 days, to ensure complete digestion of the substrate. Conversely, inadequate retention time can result in incomplete biogas recovery and system inefficiencies. The carbon-to-nitrogen (C/N) ratio of food waste is a pivotal factor in determining the efficiency of biogas recovery. A low C/N ratio can lead to excessive ammonia production, which can, in turn, inhibit microbial activity within the system. The integration of food waste with carbon-rich materials, such as agricultural residues, has been demonstrated to enhance the C/N ratio, thereby stabilizing the digestion process (Mirmohamadsadeghi et al., 2019).

Through the integration of sophisticated pretreatment methodologies with meticulously refined process parameters, researchers and practitioners have achieved a substantial enhancement in the efficiency and scalability of food waste biotransformation systems (Table 4.1).

### 4.2.2 Integration with the Circular Economy

The circular economy is a framework designed to minimize waste and maximize resource efficiency by reusing, recycling, and repurposing materials. Biotransformation systems are in alignment with this philosophy, as they convert food waste into two valuable outputs: biogas and digestate.

The utilization of biogas as a substitute for fossil fuels in electricity generation, heating, and transportation is a viable solution. A distinguishing feature of biogas is its carbon-neutrality, which is achieved through a process of offsetting the released carbon dioxide ($CO_2$) with that absorbed by the plants and organic materials that initially generated the food waste. This process ensures a net zero emission of $CO_2$, making biogas a sustainable energy option. This characteristic of biogas renders it an environmentally friendly alternative for reducing greenhouse gas emissions (Chew et al., 2021).

The nutrient-rich by-product of anaerobic digestion, digestate, contains high levels of nitrogen, phosphorus, and potassium—key nutrients for plant growth. The utilization of digestate as a fertilizer reduces the reliance on synthetic fertilizers, thereby promoting sustainable agricultural practices. Furthermore, the utilization of digestate has been shown to enhance soil health by improving its structure and nutrient content (Ferdeş et al., 2022).

The diversion of food waste from landfills to biotransformation systems has been demonstrated to result in a substantial reduction in methane emissions. This is due to the direct release of methane into the atmosphere during landfill decomposition. Methane, a highly potent greenhouse gas, possesses a global warming potential that far exceeds that

**Table 4.1** Recent studies on the pretreatment of food waste for biogas production

| Pretreatment technique | Description | Example of food waste | Effect on biotransformation | Reference |
|---|---|---|---|---|
| Physical pretreatment | Mechanical processes like grinding, chopping, or heating | Fresh and dry leaves (samples of park wastes) | Increases surface area for microbial digestion; improves hydrolysis and subsequent methane production | Ali and Sun (2015) |
| Thermal pretreatment | Heating food waste to 130 °C for 50 min to enhance solubilization | Food waste mixtures | COD solubilization increased by 47%, significantly enhancing biogas yield | Menon et al. (2016) |
| Chemical pretreatment | Using alkalis or acids to degrade lignocellulosic components | Coffee production waste | Alkali pretreatment with NaOH improved hydrolysis and boosted methane production by up to 83% | Battista et al. (2016) |
| Alkali pretreatment | Applying 1% CaO to improve solubilization and methane yield | Starch-heavy waste (e.g., rice, bread) | Enhanced biogas production by 65.48% methane yield, stabilizing organic material breakdown | Linyi et al. (2020) |
| Biological pretreatment | Using lignin-degrading fungal strains like *Pleurotus ostreatus* | Municipal solid waste | Increased solubilization of lignocellulosic materials, leading to a 169.5% increase in biogas production | Bala and Mondal (2020) |
| Enzymatic pretreatment | Addition of enzymes (rumen fluid) to hydrolyze cellulose into fermentable sugars | Grass and sewage sludge | Enhanced hydrolysis efficiency, improving digestibility and increasing methane production | Hren et al. (2020) |

of carbon dioxide. Anaerobic digestion systems that capture and utilize methane have been developed as a means of mitigating environmental harm while generating usable energy.

Furthermore, life cycle assessments (LCAs) have repeatedly demonstrated that biotransformation systems exhibit a reduced environmental impact in comparison to alternative waste management methodologies, such as incineration or landfilling. Additionally, they have been shown to exhibit a positive energy balance, indicating that the energy produced through biogas generation exceeds the energy necessary for system operation (Hadidi et al., 2023).

The integration of biotransformation systems into urban environments has gained significant traction in recent years, with the aim of addressing the pressing challenges of localized waste management. Decentralized anaerobic digestion facilities are being implemented in various locations, including communities, restaurants, and food processing facilities, to process food waste in close proximity to its source. This approach has the potential to reduce transportation-related emissions and costs, while concurrently providing local energy solutions. Urban biotransformation systems also create opportunities for community engagement and education, fostering sustainable practices at the grassroots level.

### 4.2.3 Emerging Technologies and Innovations

Thermophilic digestion functions at elevated temperatures (50–60 °C) in contrast to the typical range of mesophilic digestion (30–40 °C), exhibiting numerous benefits. The elevated temperatures accelerate microbial activity, reduce pathogens, and increase methane production rates. Thermophilic systems have been shown to be particularly effective for food waste with high fat and protein content, as these compounds exhibit enhanced degradation under high-temperature conditions. Research findings indicate that methane content can reach 54–58% under thermophilic conditions, making it a preferred choice for industrial-scale applications (Tsavkelova et al., 2012).

The field of nanotechnology has the potential to transform the biotransformation of food waste. The employment of nanocatalysts, including iron oxide nanoparticles, has been demonstrated to enhance the molecular-level breakdown of organic material. These catalysts have been shown to facilitate chemical reactions, thereby enhancing the speed and efficiency of anaerobic digestion. For instance, studies have demonstrated that nanocatalyzed systems can enhance methane production by up to 23.5% in comparison to conventional methods (Bharathi et al., 2022).

Automation and digital monitoring systems have been demonstrated to play a critical role in improving the reliability and efficiency of biotransformation processes. Sensors and software meticulously monitor parameters such as temperature, pH, and organic loading rates in real-time, facilitating precise adjustments to ensure optimal conditions are maintained. The implementation of these technologies has been shown to result in a reduction of operational costs and an elimination of downtime, thereby rendering anaerobic digestion systems more viable for large-scale and industrial applications (Chew et al., 2021).

Emerging technologies continue to address key challenges in biotransformation, such as scalability, process stability, and environmental impact. Innovations in thermophilic systems, nanotechnology, and automation are paving the way for more efficient, sustainable, and accessible solutions. Continued research and development in these areas is expected to position these technologies as crucial components in global initiatives to transition to renewable energy sources and to reduce waste.

The transformation of food waste into biogas represents a pivotal aspect of sustainable waste management and renewable energy production. Significant advancements have been made in pretreatment techniques, the optimization of anaerobic digestion parameters, and the integration of these systems into the circular economy, resulting in substantial improvements in the efficiency and scalability of biogas production. The potential of biotransformation to address pressing global challenges is further enhanced by emerging technologies, such as thermophilic digestion, nanotechnology, and automation.

## 4.3  Reactor Configuration and Design

Biogas-producing bioreactors play a pivotal role in the sustainable management of organic waste, including food waste, by converting it into renewable energy and valuable by-products (Deepanraj et al., 2015). These reactors employ the process of anaerobic digestion (AD), wherein microorganisms decompose organic material in the absence of oxygen, yielding biogas, which is predominantly methane ($CH_4$) and carbon dioxide ($CO_2$) (Economou et al., 2024). The biogas produced can be utilized as a clean energy source for electricity generation, heating, and vehicle fuel (Liu et al., 2017). The residual material, known as digestate, serves as a nutrient-rich fertilizer (Zhou et al., 2019).

The efficiency of anaerobic digestion is contingent upon the design and configuration of the bioreactor. The biogas yield and quality are influenced by various factors, including substrate composition, reactor type, temperature, pH, and microbial activity (Yeshanew et al., 2016). To accommodate diverse substrate types and operational needs, a range of bioreactors have been developed, including batch reactors, continuous stirred tank reactors (CSTRs), fixed-bed reactors, and multi-stage systems (Ahamed et al., 2015). Recent trends in reactor design have focused on addressing challenges such as process stability, methane enrichment, and scalability (Petracchini et al., 2018).

Advancements in bioreactor configuration have enabled the attainment of higher methane yields, more stable operations, and improved adaptability to diverse feedstocks, such as food waste (Zhou et al., 2019). Innovative designs, such as multi-phased systems, bioelectrochemical integration, and compact multi-stage configurations, enhance the efficiency and environmental sustainability of biogas production (Liu et al., 2017).

### 4.3.1 Multi-phased and Staged Reactors

Multi-phased and staged reactors are among the most significant advancements in anaerobic digestion technology. Unlike traditional single-stage reactors, which process all stages of digestion within one-unit, multi-phased systems separate the hydrolysis/acidogenesis and methanogenesis phases. This separation allows each phase to operate under optimal conditions, enhancing process stability and biogas yields. Hydrolysis and acidogenesis, dominated by acid-forming bacteria, require acidic conditions to break down complex organic materials into volatile fatty acids (VFAs). Methanogenesis, in contrast, depends on methanogenic archaea that thrive in neutral to slightly alkaline environments.

One successful application of this approach is the Multi-Phased Anaerobic Baffled Reactor (MP-ABR). In an MP-ABR, food waste passes through multiple chambers, each designed to support a specific stage of digestion. A study demonstrated that a four-chambered MP-ABR operating at 35 °C achieved an 85.3% reduction in chemical oxygen demand (COD) and biogas production of 215.57 mL per gram of volatile solids removed daily. The reactor's baffled design prevented VFA accumulation in the methanogenic phase, stabilizing methane production (Ahamed et al., 2015) (Fig. 4.2).

The study by Van Hecke et al. (2019) demonstrated the effectiveness of a two-stage anaerobic digestion system that separated hydrolysis and methanogenesis into individual tanks. This design improved process stability by reducing acid accumulation, a common issue in single-stage reactors that can inhibit methanogenic activity. By creating optimal conditions for both phases, the system enhanced microbial efficiency and methane production. Compared to single-stage reactors, this approach achieved a 20% increase in methane yields, highlighting its potential for improving biogas production from complex substrates like food waste. The study underscores the benefits of phase separation in anaerobic digestion processes.

Compact multi-stage anaerobic digesters segment the digestion process into separate stages, each optimized for a specific phase of anaerobic digestion. These reactors aim to reduce the physical footprint of biogas systems while enhancing efficiency and methane yields. For example, the first stage may focus on hydrolysis and acidogenesis, while the second stage targets methanogenesis. A compact multi-stage anaerobic digestion reactor was designed for urban food waste management. This reactor achieved methane concentrations of 85%, demonstrating high efficiency in biogas production. A key innovation was the use of digestate recirculation, which stabilized pH levels and prevented ammonia inhibition, ensuring optimal microbial activity. The compact design addressed space

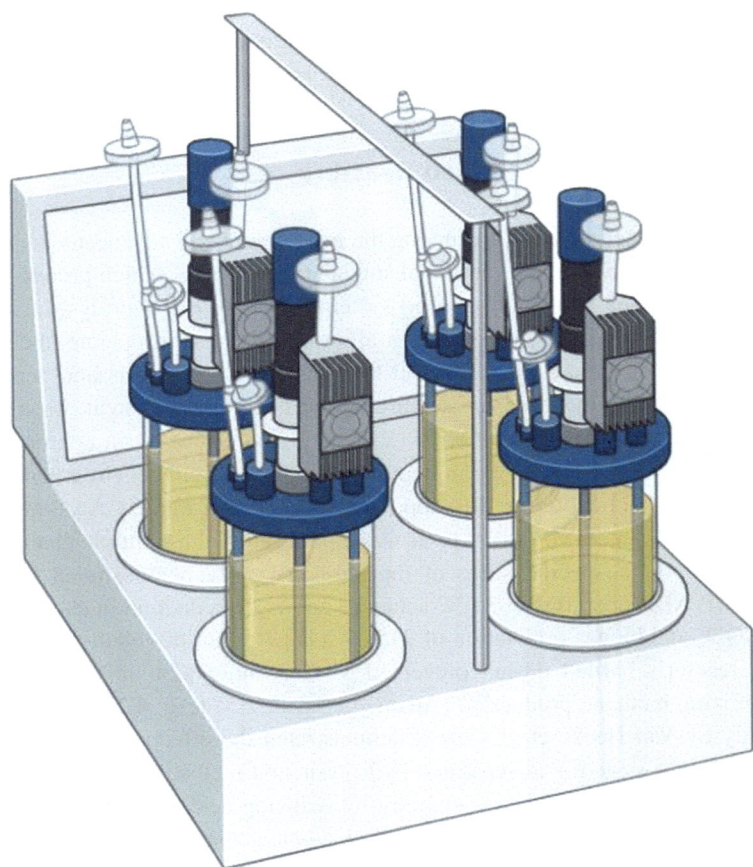

**Fig. 4.2** Multi-phased baffled reactors

constraints typical of urban environments, making it suitable for decentralized waste management. The reactor's modular configuration and operational stability highlighted its potential for scalable applications in urban areas, providing an effective solution for sustainable food waste management and renewable energy generation (Petracchini et al., 2018).

Multi-phased systems are particularly effective for high-solid substrates such as food waste, which undergoes rapid acidogenesis. By isolating phases, these reactors reduce the risk of methanogen inhibition, increase process efficiency, and support higher biogas yields. They are ideal for municipal and industrial applications, where food waste often exhibits high variability in composition and organic load.

### 4.3.2 Bioelectrochemical Systems (BES) Integration

Bioelectrochemical systems (BES) are a cutting-edge addition to anaerobic digestion technology. These systems incorporate electrodes into reactors to facilitate microbial activity, accelerating the degradation of organic matter and enhancing methane production. BES reactors promote direct electron transfer between electroactive microbes and electrodes, improving the conversion efficiency of organic substrates into biogas. Additionally, they allow for methane enrichment, increasing the proportion of methane in the produced biogas.

An example of BES integration is a thermophilic BES reactor treating food waste, which achieved methane enrichment up to 98%. In this system, a two-chamber reactor separated oxidation and reduction reactions, creating an environment that favored methanogenic archaea while inhibiting competing microbial pathways. This improvement was attributed to the enhanced electron transfer facilitated by the BES, which accelerated microbial activity and organic matter degradation. However, the study also highlighted challenges such as the accumulation of volatile fatty acids and high operational costs associated with electrode materials (Liu et al., 2017).

Shen et al. (2021) explored the use of biochar as an electrode material in BES reactors to address the high costs associated with conventional electrode materials like graphite or platinum. Biochar, derived from biomass, is a sustainable and low-cost alternative. The study found that BES reactors using biochar electrodes achieved methane yields comparable to those using traditional materials. This finding demonstrates the feasibility of biochar in reducing implementation costs without compromising reactor performance. By using renewable electrode materials, this innovation makes BES technology more accessible for large-scale adoption, particularly in regions where cost constraints limit the feasibility of advanced waste management systems.

Despite these challenges, BES reactors hold great promise for improving biogas quality and process efficiency. With further optimization, such as reducing material costs and mitigating VFA accumulation, BES technology could play a transformative role in large-scale biogas systems, particularly in settings where high-purity methane is required.

### 4.3.3 Batch and Continuous Systems

The operational mode of a reactor—batch or continuous—significantly impacts its efficiency and suitability for specific applications (Fig. 4.3). Batch reactors operate by loading a fixed amount of substrate, allowing complete digestion, and then emptying the reactor before restarting the cycle. These systems are simple to construct and operate, making them ideal for small-scale or decentralized settings. However, their efficiency is limited by downtime between cycles, and they are less suitable for large-scale operations.

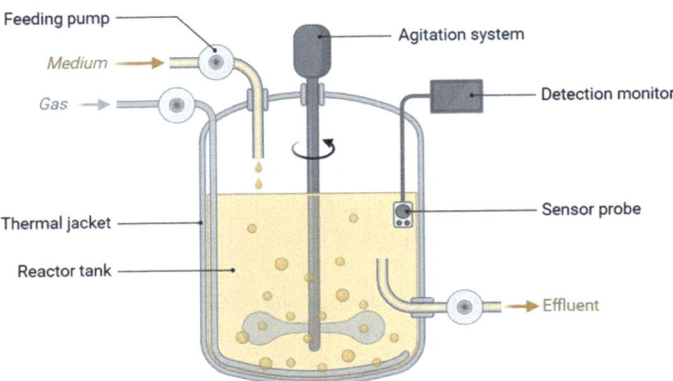

**Fig. 4.3** Exemplification of a stirred tank reactor (batch or continuous system)

Continuous reactors, such as Continuous Stirred Tank Reactors (CSTRs), operate by continuously feeding substrate and removing digestate (Fig. 4.3). This steady-state operation ensures consistent biogas production and enables higher throughput. CSTRs are widely used for industrial applications due to their scalability and ability to handle large volumes of food waste. For instance, a 300 m$^3$ CSTR implemented at a Chinese university processed 5 tons of food and garden waste daily, producing 144,000 m$^3$ of biogas annually. This system powered campus facilities and demonstrated the integration of biogas technology into institutional operations (Zhou et al., 2019).

Household-scale batch digesters implemented in rural India demonstrated their effectiveness in converting food and agricultural waste into biogas, primarily used for cooking fuel. These digesters were praised for their simplicity, low cost, and suitability for decentralized applications. The study highlighted how these systems provide a reliable energy source for rural households, reducing dependency on traditional fuels such as firewood or kerosene. By diverting organic waste to biogas production, these digesters also contributed to better waste management and reduced greenhouse gas emissions. Furthermore, the digestate produced was used as organic fertilizer, supporting local agriculture and promoting sustainability (Säf, 2010).

The hybrid batch-continuous system combined batch feeding with continuous digestate removal, striking a balance between simplicity and operational stability. This innovative configuration minimized downtime typically associated with batch systems while maintaining the steady biogas production characteristic of continuous systems. The study found that this approach stabilized methane yields and prevented process disruptions caused by substrate variability. Such a system proved particularly effective for small to medium-scale applications where complete reliance on either batch or continuous systems might not be practical, offering a flexible and efficient solution for food waste digestion (Deepanraj et al., 2015).

These studies collectively highlight the versatility of biogas technologies, ranging from household applications to large-scale institutional and hybrid systems, addressing diverse energy and waste management needs. While batch systems remain relevant for small-scale applications, continuous systems are better suited for industrial-scale operations due to their higher efficiency, scalability, and ability to handle variations in feedstock composition.

### 4.3.4 Temperature Optimized Bioreactors

Temperature is a critical factor influencing microbial activity and biogas production rates in anaerobic digestion. Two primary temperature regimes are used: mesophilic (35–40 °C) and thermophilic (50–60 °C). Mesophilic systems are easier to maintain and offer greater stability, making them suitable for a wide range of applications. However, thermophilic systems provide faster reaction rates, higher methane yields, and better pathogen reduction, albeit at the cost of increased energy requirements for heating and greater sensitivity to operational fluctuations.

Thermophilic reactors have shown particular promise for treating food waste. In one study, a thermophilic batch reactor achieved biogas yields of 7556 mL from food waste at 50 °C, significantly outperforming mesophilic systems. The higher temperatures accelerated the breakdown of organic materials, particularly fats and proteins, resulting in increased methane production (Deepanraj et al., 2015).

The integration of heat exchangers in mesophilic reactors demonstrated an 18% improvement in methane yields. By maintaining consistent temperature conditions, the system enhanced microbial activity and prevented temperature fluctuations that could inhibit digestion processes. The study validated the effectiveness of heat exchangers in stabilizing reactor performance and improving efficiency in mesophilic systems, which are known for their cost-effectiveness and operational stability (Riggio et al., 2017).

Ventura et al. (2014) explored the benefits of alternating between mesophilic (35–40 °C) and thermophilic (50–60 °C) conditions in anaerobic digestion. The dual-temperature reactors efficiently processed food and yard waste, balancing the rapid reaction rates of thermophilic regimes with the stability of mesophilic conditions. This approach maximized methane yields while optimizing energy input, making it a practical solution for facilities with variable feedstocks or energy limitations.

Thermophilic systems offer higher yields but require greater energy input, while innovations like heat exchangers and dual-temperature regimes enhance the efficiency and adaptability of mesophilic systems. Modern reactor designs increasingly incorporate temperature control systems, such as heat exchangers and insulated chambers, to maintain consistent operating temperatures. These innovations enhance process reliability and ensure optimal microbial performance.

## 4.3.5 Fixed-Bed and Packed-Bed Reactors

Fixed-bed reactors are reactors where the catalyst or biomass-supporting material remains stationary within the reactor, while the reactants flow through the bed. These reactors are widely used in industries ranging from petrochemical to renewable energy due to their simple design and efficient reaction processes. In contrast, packed-bed reactors, a subset of fixed-bed reactors, specifically use a bed packed with solid materials like catalytic particles or biofilm carriers. These materials provide a large surface area for reactions or microbial growth, making packed-bed reactors especially suitable for biogas production and gas–liquid reactions (Rodríguez-Durán et al., 2023) (Fig. 4.4).

Fixed-bed and packed-bed reactors are gaining popularity for their ability to retain microbial biomass within the reactor. Packed-bed reactors often utilize biofilm supports like activated carbon, expanded clay, or plastic rings to enhance reaction rates and optimize gas–liquid contact. In fixed-bed designs, the flow can be arranged as downward or upward, depending on the process requirements, ensuring maximum interaction between the gases and catalyst or biomass. Both reactor types are pivotal in processes like anaerobic digestion, methane reforming, and hydrogen production due to their adaptability and high efficiency.

One significant trend is the focus on enhancing biogas yield. Fixed-bed reactors, for instance, have shown great potential in flexible, demand-driven biogas production, particularly under thermophilic conditions that improve methane yield and response to variable feed patterns (Terboven et al., 2017).

In hydrogen production, packed-bed reactors have been equipped with innovative biofilm supports like activated carbon, which facilitate biohydrogen generation while minimizing methane emissions (Chang et al., 2002). These reactors also exhibit versatility, as seen in dual-purpose designs capable of biogas scrubbing (e.g., $H_2S$ removal) and methane-dependent water quality improvements, thus serving multiple roles in renewable energy systems (Tanaka, 2002).

Advancements in heat and flow management also feature prominently in recent developments. Packed-bed reactors now incorporate better heat management systems, enhancing their efficiency for high-temperature processes like methane reforming. These innovations allow for better reaction rates and improved scalability for industrial applications (Spallina et al., 2013).

Fixed-bed and packed-bed reactors are critical technologies in the field of biogas production, offering adaptability, scalability, and high efficiency. Ongoing innovations ensure these reactors remain at the forefront of renewable energy and biochemical engineering, addressing challenges in energy production and environmental sustainability.

4.3 Reactor Configuration and Design

**Fig. 4.4** Schematic representation of a packed bed reactor

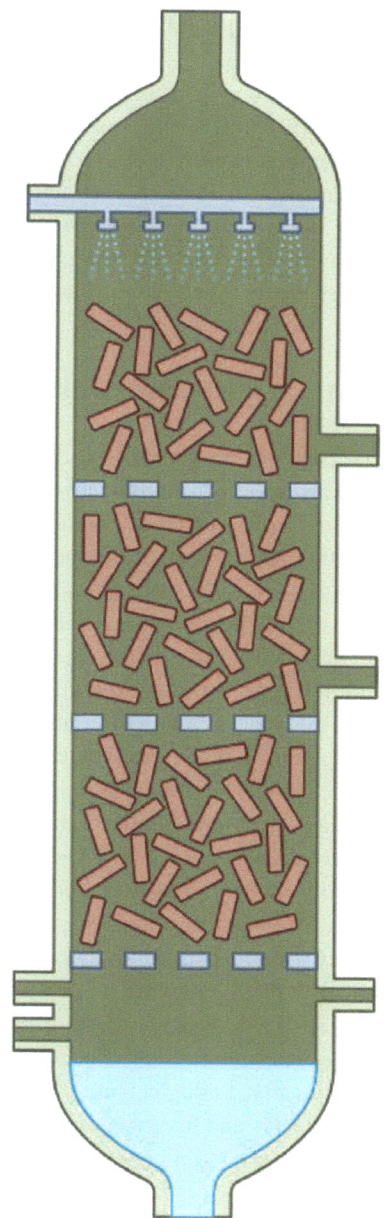

## 4.4 Biogas Utilization and Energy Recovery

Biogas has emerged as a vital renewable energy source due to its potential to address energy demands while simultaneously reducing greenhouse gas (GHG) emissions and managing organic waste. Produced through anaerobic digestion (AD), biogas consists primarily of methane ($CH_4$) and carbon dioxide ($CO_2$), along with minor quantities of other gases such as hydrogen sulfide ($H_2S$) and water vapor. Recent advancements in biogas production, upgrading, and utilization have revolutionized the field, improving energy recovery, environmental sustainability, and economic feasibility.

### 4.4.1 Innovations in Biogas Production and Upgrading

The foundation of biogas production is the anaerobic digestion process, where microorganisms break down organic material in an oxygen-free environment. The efficiency of this process depends on the quality of feedstock, reactor designs, and process optimization. Advances in these areas have significantly increased biogas yields. Feedstock co-digestion, for instance, combines different organic materials such as agricultural waste, municipal solid waste, and industrial byproducts, enhancing microbial activity and biogas production (Pöschl et al., 2010). Modern reactor designs, including vertical stirred tank reactors and plug-flow digesters, ensure optimal conditions for microbial digestion, reducing energy input requirements while increasing methane output.

Upgrading biogas to biomethane, a high-purity form of methane suitable for diverse applications, is a critical step in expanding its utility. Traditional upgrading technologies, such as water scrubbing and amine scrubbing, have been refined to improve performance. Water scrubbing exploits $CO_2$'s higher solubility in water, offering an environmentally friendly method for purification, though it requires significant water resources (Vilardi et al., 2020). Amine scrubbing, which uses chemical solvents to selectively absorb $CO_2$, achieves high methane recovery but has high energy demands (Wu et al., 2016). More recently, pressure swing adsorption (PSA) has emerged as a leading technique. Innovations such as the use of metal–organic frameworks (MOFs) in PSA systems have reduced energy consumption by over 50% compared to traditional adsorbents (Wu et al., 2015). Additionally, biological methanation, an emerging technology, uses microorganisms to convert $CO_2$ and hydrogen into methane, offering a cost-effective and sustainable solution for biogas upgrading (Angelidaki et al., 2018).

### 4.4.2 Recovery and Utilization Pathways

To maximize energy recovery, biogas systems are increasingly integrating advanced technologies. One such advancement is the Organic Rankine Cycle (ORC), which captures and converts waste heat from biogas plants into additional electricity. This integration significantly boosts the overall energy efficiency of biogas plants, reducing thermal energy losses and increasing electricity production by up to 20% (Baccioli et al., 2019). ORC systems are particularly valuable in cold climates, where waste heat is otherwise dissipated.

Energy and exergy analyses have become essential tools for optimizing biogas utilization pathways. Exergy analysis, which evaluates the quality of energy transformations within a system, helps identify inefficiencies and potential improvements. Recent studies have shown that water scrubbing has the highest exergy efficiency (94.5%) among conventional upgrading methods, while innovations like PSA with MOFs further enhance performance (Vilardi et al., 2020). These tools guide the design and operation of biogas systems, ensuring maximum energy recovery with minimal losses.

Biogas is a versatile energy source that can be utilized in multiple ways, including heat and power generation, fuel cells, and as a transportation fuel. Each pathway offers unique advantages and challenges, tailored to specific applications.

Combined Heat and Power (CHP) systems are the most common method of biogas utilization. These systems generate electricity and capture residual heat for use in industrial or residential applications, achieving efficiencies of up to 90% (Hakawati et al., 2017). Modern CHP systems use advanced gas engines or microturbines, which are more efficient and environmentally friendly compared to older technologies. The primary challenge lies in maintenance costs and emissions control, which require constant innovation to remain competitive.

Fuel cells, particularly Molten Carbonate Fuel Cells (MCFCs) and Solid Oxide Fuel Cells (SOFCs), represent an innovative approach to biogas utilization. Operating at high temperatures, these cells exhibit high conversion efficiencies and can handle biogas impurities better than low-temperature cells (Bove & Lunghi, 2005). MCFCs, for instance, can directly utilize biogas without extensive upgrading, making them suitable for industrial applications. However, the high costs and durability issues associated with fuel cells remain significant barriers to widespread adoption.

Biomethane, derived from upgraded biogas, can be compressed (CNG) or liquefied (LNG) for use in transportation. As a clean fuel, it emits fewer GHGs compared to conventional fossil fuels. However, infrastructure challenges, such as limited refueling stations and high initial investment costs, hinder its broader adoption. Policymakers and industry leaders are working to address these barriers through subsidies and infrastructure development (Park, 2014).

### 4.4.3 Environmental, Economic, and Policy Impacts

Biogas utilization offers significant environmental benefits, primarily through the reduction of GHG emissions. By substituting biogas for fossil fuels in power generation, transportation, and heating, $CO_2$ emissions are significantly reduced. Moreover, the anaerobic digestion process mitigates methane emissions from organic waste decomposition, a major contributor to climate change (Chen et al., 2012). Life-cycle assessments reveal that the comprehensive utilization of biogas and its byproducts, such as digestate, further enhances its environmental performance. Digestate, the residue from anaerobic digestion, is rich in nutrients like nitrogen and phosphorus. Innovative processing methods, such as struvite precipitation and ammonia stripping, recover these nutrients, creating high-value fertilizers while reducing environmental pollution (Szymańska et al., 2019).

Economically, while biogas systems require substantial initial investments, they are increasingly viable due to long-term savings, government incentives, and carbon credits. Efficient designs and advanced technologies have shortened the payback periods for biogas plants. For example, optimized upgrading systems with biomethane production can achieve payback periods as short as 8.9 years (Wu et al., 2016).

Policy frameworks play a crucial role in shaping the biogas industry. In regions like the European Union, governments have introduced incentives for biogas upgrading and biomethane production. Subsidies for renewable energy projects, carbon pricing mechanisms, and mandates for GHG reduction have driven significant investments in biogas infrastructure (Park, 2014). For example, policies promoting the injection of biomethane into natural gas grids have expanded its market potential, enabling its use in diverse applications, including heating, power generation, and transportation.

In emerging markets, policymakers are focusing on cost-effective technologies to make biogas more accessible. Public–private partnerships, capacity-building programs, and research funding are accelerating the development and deployment of biogas systems. However, challenges such as inadequate infrastructure, lack of technical expertise, and policy inconsistencies must be addressed to unlock the full potential of biogas.

The future of biogas lies in the integration of advanced technologies, expansion into new markets, and stronger policy support. Continued innovation in upgrading methods, such as biological methanation, and the development of cost-effective solutions like small-scale digesters for rural areas, will drive the growth of the biogas industry. Collaboration between governments, industries, and research institutions will be essential to overcome existing barriers and establish biogas as a cornerstone of sustainable energy systems.

## 4.5 Future Trends in Biogas Utilization and Technology

Biogas is emerging as a critical component of renewable energy systems, driven by technological advancements, sustainability initiatives, and evolving energy demands. These advancements are shaping the future of biogas across various sectors, from energy production to industrial applications.

One of the most significant trends in biogas is the enhancement of upgrading technologies. Hybrid systems, combining traditional methods like pressure swing adsorption with

## 4.5 Future Trends in Biogas Utilization and Technology

membranes or cryogenic separation, are achieving higher methane purity with reduced operational costs. These innovations, along with second-generation upgrading processes that convert $CO_2$ into methane or methanol, are positioning biogas as a key player in energy storage and emissions reduction. These advancements not only improve system efficiency but also support the broader integration of biogas into renewable energy networks (Assunção et al., 2021).

Biogas is also being integrated with renewable energy systems to address energy intermittency. Power-to-gas technologies enable the conversion of surplus electricity into methane, providing a flexible energy carrier for long-term storage and grid balancing. This integration enhances the reliability of renewable energy systems and makes biogas a critical component of future energy infrastructures (Angelidaki et al., 2018).

Small-scale and decentralized biogas systems are gaining traction in rural and off-grid areas, addressing both energy needs and waste management challenges. These systems are particularly impactful in developing regions, where they provide access to renewable energy and sustainable waste solutions. In urban settings, centralized anaerobic digestion facilities are optimizing large-scale waste management and energy production, demonstrating the versatility of biogas across different environments (Maurya et al., 2019).

Industrial applications of biogas are expanding rapidly. Solid oxide fuel cells (SOFCs), known for their high efficiency and ability to tolerate biogas impurities, are increasingly being used in industrial energy systems. Additionally, biogas is being explored as a sustainable feedstock for chemical production, including methanol and hydrogen, aligning with global decarbonization efforts. These applications highlight the potential of biogas to transform industrial energy use and contribute to a low-carbon economy (Thiruselvi et al., 2021).

Policy and market trends are further accelerating biogas adoption. Governments in regions like the European Union are promoting biomethane injection into natural gas grids and offering tax incentives for renewable energy projects. These initiatives align with global sustainability goals, such as achieving net-zero emissions, and emphasize the role of biogas in a circular bioeconomy (Kougias & Angelidaki, 2018).

Efforts to reduce costs and enhance efficiency are also advancing biogas systems. Innovations such as enzymatic hydrolysis and thermochemical pretreatments are improving substrate breakdown, increasing biogas yields, and reducing energy consumption (Bala et al., 2019). These cost-effective measures are making biogas systems more competitive and accessible.

In summary, the future of biogas is marked by technological innovation, integration with renewable systems, and broader adoption across industrial and rural settings. As supportive policies and market incentives continue to grow, biogas is positioned to play a central role in global energy transition efforts, contributing to a sustainable and low-carbon future.

## References

Ahamed, A., Chen, C.-L., Rajagopal, R., Wu, D., Mao, Y., Ho, I. J. R., Lim, J. W., & Wang, J.-Y. (2015). Multi-phased anaerobic baffled reactor treating food waste. *Bioresource Technology, 182*, 239–244. https://doi.org/10.1016/j.biortech.2015.01.117

Al-Wahaibi, A., Osman, A. I., Al-Muhtaseb, A. H., Alqaisi, O., Baawain, M., Fawzy, S., & Rooney, D. W. (2020). Techno-economic evaluation of biogas production from food waste via anaerobic digestion. *Scientific Reports, 10*(1), 15719. https://doi.org/10.1038/s41598-020-72897-5

Alghoul, O., El-Hassan, Z., Ramadan, M., & Olabi, A. G. (2019). Experimental investigation on the production of biogas from waste food. *Energy Sources, Part a: Recovery, Utilization, and Environmental Effects, 41*(17), 2051–2060. https://doi.org/10.1080/15567036.2018.1549156

Ali, S. S., & Sun, J. (2015). Physico-chemical pretreatment and fungal biotreatment for park wastes and cattle dung for biogas production. *Springerplus, 4*(1), 712. https://doi.org/10.1186/s40064-015-1466-9

Angelidaki, I., Treu, L., Tsapekos, P., Luo, G., Campanaro, S., Wenzel, H., & Kougias, P. G. (2018). Biogas upgrading and utilization: Current status and perspectives. *Biotechnology Advances, 36*(2), 452–466. https://doi.org/10.1016/j.biotechadv.2018.01.011

Assunção, L. R. C., Mendes, P. A. S., Matos, S., & Borschiver, S. (2021). Technology roadmap of renewable natural gas: Identifying trends for research and development to improve biogas upgrading technology management. *Applied Energy, 292*, 116849. https://doi.org/10.1016/j.apenergy.2021.116849

Baccioli, A., Ferrari, L., Vizza, F., & Desideri, U. (2019). Potential energy recovery by integrating an ORC in a biogas plant. *Applied Energy, 256*, 113960. https://doi.org/10.1016/j.apenergy.2019.113960

Bala, R., & Mondal, M. K. (2020). Study of biological and thermo-chemical pretreatment of organic fraction of municipal solid waste for enhanced biogas yield. *Environmental Science and Pollution Research, 27*(22), 27293–27304. https://doi.org/10.1007/s11356-019-05695-w

Bala, R., Gautam, V., & Mondal, M. K. (2019). Improved biogas yield from organic fraction of municipal solid waste as preliminary step for fuel cell technology and hydrogen generation. *International Journal of Hydrogen Energy, 44*(1), 164–173. https://doi.org/10.1016/j.ijhydene.2018.02.072

Battista, F., Fino, D., & Mancini, G. (2016). Optimization of biogas production from coffee production waste. *Bioresource Technology, 200*, 884–890. https://doi.org/10.1016/j.biortech.2015.11.020

Bharathi, P., Dayana, R., Rangaraju, M., Varsha, V., Subathra, M., Gayathri, & Sundramurthy, V. P. (2022). Biogas Production from Food Waste Using Nanocatalyst. *Journal of Nanomaterials, 2022*(1), 7529036. https://doi.org/10.1155/2022/7529036

Bove, R., & Lunghi, P. (2005). Experimental comparison of MCFC performance using three different biogas types and methane. *Journal of Power Sources, 145*(2), 588–593. https://doi.org/10.1016/j.jpowsour.2005.01.069

Chang, J.-S., Lee, K.-S., & Lin, P.-J. (2002). Biohydrogen production with fixed-bed bioreactors. *International Journal of Hydrogen Energy, 27*(11), 1167–1174. https://doi.org/10.1016/S0360-3199(02)00130-1

Chen, S., Chen, B., & Song, D. (2012). Life-cycle energy production and emissions mitigation by comprehensive biogas–digestate utilization. *Bioresource Technology, 114*, 357–364. https://doi.org/10.1016/j.biortech.2012.03.084

Chew, K. R., Leong, H. Y., Khoo, K. S., Vo, D.-V.N., Anjum, H., Chang, C.-K., & Show, P. L. (2021). Effects of anaerobic digestion of food waste on biogas production and environmental impacts:

A review. *Environmental Chemistry Letters, 19*(4), 2921–2939. https://doi.org/10.1007/s10311-021-01220-z

Deepanraj, B., Sivasubramanian, V., & Jayaraj, S. (2015). Kinetic study on the effect of temperature on biogas production using a lab scale batch reactor. *Ecotoxicology and Environmental Safety, 121*, 100–104. https://doi.org/10.1016/j.ecoenv.2015.04.051

Economou, F., Voukkali, I., Papamichael, I., Phinikettou, V., Loizia, P., Naddeo, V., Sospiro, P., Liscio, M. C., Zoumides, C., Țîrcă, D. M., & Zorpas, A. A. (2024). Turning food loss and food waste into watts: A review of food waste as an energy source. *Energies, 17*(13), 1–30. https://doi.org/10.3390/en17133191

Ferdeș, M., Zăbavă, B. Ș., Paraschiv, G., Ionescu, M., Dincă, M. N., & Moiceanu, G. (2022). Food waste management for biogas production in the context of sustainable development. *Energies, 15*(17). https://doi.org/10.3390/en15176268

Hadidi, M., Bahlaouan, B., Antri, S. E., Benali, M., & Boutaleb, N. (2023). Biotransformation of food waste to bio-products: Biogas and biofertilizer. *International Journal of Environmental Studies, 80*(3), 672–686. https://doi.org/10.1080/00207233.2022.2096953

Hakawati, R., Smyth, B. M., McCullough, G., De Rosa, F., & Rooney, D. (2017). What is the most energy efficient route for biogas utilization: Heat, electricity or transport? *Applied Energy, 206*, 1076–1087. https://doi.org/10.1016/j.apenergy.2017.08.068

Hren, R., Petrovič, A., Čuček, L., & Simonič, M. (2020). Determination of various parameters during thermal and biological pretreatment of waste materials. *Energies, 13*(9). https://doi.org/10.3390/en13092262

Jin, Y., Chen, T., Chen, X., & Yu, Z. (2015). Life-cycle assessment of energy consumption and environmental impact of an integrated food waste-based biogas plant. *Applied Energy, 151*, 227–236. https://doi.org/10.1016/j.apenergy.2015.04.058

Kougias, P. G., & Angelidaki, I. (2018). Biogas and its opportunities—A review. *Frontiers of Environmental Science & Engineering, 12*(3), 14. https://doi.org/10.1007/s11783-018-1037-8

Linyi, C., Yujie, Q., Buqing, C., Chenglong, W., Shaohong, Z., Renglu, C., Shaohua, Y., Lan, Y., & Zhiju, L. (2020). Enhancing degradation and biogas production during anaerobic digestion of food waste using alkali pretreatment. *Environmental Research, 188*, 109743. https://doi.org/10.1016/j.envres.2020.109743

Liu, S. Y., Charles, W., Ho, G., Cord-Ruwisch, R., & Cheng, K. Y. (2017). Bioelectrochemical enhancement of anaerobic digestion: Comparing single- and two-chamber reactor configurations at thermophilic conditions. *Bioresource Technology, 245*, 1168–1175. https://doi.org/10.1016/j.biortech.2017.08.095

Maurya, R., Tirkey, S. R., Rajapitamahuni, S., Ghosh, A., & Mishra, S. (2019). Chapter 9—Recent advances and future prospective of biogas production. In B. Hosseini (Ed.), *Woodhead publishing series in energy* (pp. 159–178). Woodhead Publishing. https://doi.org/10.1016/B978-0-12-817937-6.00009-6

Menon, A., Ren, F., Wang, J.-Y., & Giannis, A. (2016). Effect of pretreatment techniques on food waste solubilization and biogas production during thermophilic batch anaerobic digestion. *Journal of Material Cycles and Waste Management, 18*(2), 222–230. https://doi.org/10.1007/s10163-015-0395-6

Mirmohamadsadeghi, S., Karimi, K., Tabatabaei, M., & Aghbashlo, M. (2019). Biogas production from food wastes: A review on recent developments and future perspectives. *Bioresource Technology Reports, 7*, 100202. https://doi.org/10.1016/j.biteb.2019.100202

Park, S. (2014). Energy recovery of organic waste: Biogas upgrading technology and policy trends. *Journal of Korea Society of Waste Management, 31*(4), 366–374.

Petracchini, F., Liotta, F., Paolini, V., Perilli, M., Cerioni, D., Gallucci, F., Carnevale, M., & Bencini, A. (2018). A novel pilot scale multistage semidry anaerobic digestion reactor to treat food waste and cow manure. *International Journal of Environmental Science and Technology, 15*(9), 1999–2008. https://doi.org/10.1007/s13762-017-1572-z

Pöschl, M., Ward, S., & Owende, P. (2010). Evaluation of energy efficiency of various biogas production and utilization pathways. *Applied Energy, 87*(11), 3305–3321. https://doi.org/10.1016/j.apenergy.2010.05.011

Pramanik, S. K., Suja, F. B., Zain, S. M., & Pramanik, B. K. (2019). The anaerobic digestion process of biogas production from food waste: Prospects and constraints. *Bioresource Technology Reports, 8*, 100310. https://doi.org/10.1016/j.biteb.2019.100310

Riggio, S., Hernandéz-Shek, M. A., Torrijos, M., Vives, G., Esposito, G., van Hullebusch, E. D., Steyer, J. P., & Escudié, R. (2017). Comparison of the mesophilic and thermophilic anaerobic digestion of spent cow bedding in leach-bed reactors. *Bioresource Technology, 234*, 466–471. https://doi.org/10.1016/j.biortech.2017.02.056

Rodríguez-Durán, L. V, Michel, M. R., Pichardo, A., & Aguilar-Zárate, P. (2023). Microbial bioreactors for secondary metabolite production. In *Microbial bioreactors for industrial molecules* (pp. 275–296). https://doi.org/10.1002/9781119874096.ch13

Säf, S. (2010). *Household biogas systems in low income rural regions*. Loughborough University.

Shen, Y., Yu, Y., Zhang, Y., Urgun-Demirtas, M., Yuan, H., Zhu, N., & Dai, X. (2021). Role of redox-active biochar with distinctive electrochemical properties to promote methane production in anaerobic digestion of waste activated sludge. *Journal of Cleaner Production, 278*, 123212. https://doi.org/10.1016/j.jclepro.2020.123212

Spallina, V., Gallucci, F., Romano, M. C., Chiesa, P., Lozza, G., & van Sint Annaland, M. (2013). Investigation of heat management for CLC of syngas in packed bed reactors. *Chemical Engineering Journal, 225*, 174–191. https://doi.org/10.1016/j.cej.2013.03.054

Szymańska, M., Szara, E., Sosulski, T., Wąs, A., Van Pruissen, G. W. P., Cornelissen, R. L., Borowik, M., & Konkol, M. (2019). A bio-refinery concept for N and P recovery—A chance for biogas plant development. *Energies, 12*(1). https://doi.org/10.3390/en12010155

Tanaka, Y. (2002). A dual purpose packed-bed reactor for biogas scrubbing and methane-dependent water quality improvement applying to a wastewater treatment system consisting of UASB reactor and trickling filter. *Bioresource Technology, 84*(1), 21–28. https://doi.org/10.1016/S0960-8524(02)00031-7

Terboven, C., Ramm, P., & Herrmann, C. (2017). Demand-driven biogas production from sugar beet silage in a novel fixed bed disc reactor under mesophilic and thermophilic conditions. *Bioresource Technology, 241*, 582–592. https://doi.org/10.1016/j.biortech.2017.05.150

Thiruselvi, D., Kumar, P. S., Kumar, M. A., Lay, C.-H., Aathika, S., Mani, Y., Jagadiswary, D., Dhanasekaran, A., Shanmugam, P., Sivanesan, S., & Show, P.-L. (2021). A critical review on global trends in biogas scenario with its up-gradation techniques for fuel cell and future perspectives. *International Journal of Hydrogen Energy, 46*(31), 16734–16750. https://doi.org/10.1016/j.ijhydene.2020.10.023

Tsavkelova, E. A., Egorova, M. A., Petrova, E. V., & Netrusov, A. I. (2012). Biogas production by microbial communities via decomposition of cellulose and food waste. *Applied Biochemistry and Microbiology, 48*(4), 377–384. https://doi.org/10.1134/S0003683812040126

Van Hecke, W., Bockrath, R., & De Wever, H. (2019). Effects of moderately elevated pressure on gas fermentation processes. *Bioresource Technology, 293*, 122129. https://doi.org/10.1016/J.BIORTECH.2019.122129

Ventura, J.-R. S., Lee, J., & Jahng, D. (2014). A comparative study on the alternating mesophilic and thermophilic two-stage anaerobic digestion of food waste. *Journal of Environmental Sciences, 26*(6), 1274–1283. https://doi.org/10.1016/S1001-0742(13)60599-9

# References

Vilardi, G., Bassano, C., Deiana, P., & Verdone, N. (2020). Exergy and energy analysis of three biogas upgrading processes. *Energy Conversion and Management, 224*, 113323. https://doi.org/10.1016/j.enconman.2020.113323

Weiland, P. (2010). Biogas production: Current state and perspectives. *Applied Microbiology and Biotechnology, 85*(4), 849–860. https://doi.org/10.1007/s00253-009-2246-7

Wu, B., Zhang, X., Xu, Y., Bao, D., & Zhang, S. (2015). Assessment of the energy consumption of the biogas upgrading process with pressure swing adsorption using novel adsorbents. *Journal of Cleaner Production, 101*, 251–261. https://doi.org/10.1016/j.jclepro.2015.03.082

Wu, B., Zhang, X., Shang, D., Bao, D., Zhang, S., & Zheng, T. (2016). Energetic-environmental-economic assessment of the biogas system with three utilization pathways: Combined heat and power, biomethane and fuel cell. *Bioresource Technology, 214*, 722–728. https://doi.org/10.1016/j.biortech.2016.05.026

Yeshanew, M. M., Frunzo, L., Pirozzi, F., Lens, P. N. L., & Esposito, G. (2016). Production of biohythane from food waste via an integrated system of continuously stirred tank and anaerobic fixed bed reactors. *Bioresource Technology, 220*, 312–322. https://doi.org/10.1016/j.biortech.2016.08.078

Zhou, M., Deng, L., Li, H., & Zou, Z. (2019). Design and efficiency analysis of biogas engineering for the mixture of kitchen waste and garden waste. In *3rd joint international information technology, mechanical and electronic engineering conference (JIMEC 2018), 3*(Jimec 2018) (pp. 42–45).